Springer Series in
OPTICAL SCIENCES
121

T0134639

Springer Series in
OPTICAL SCIENCES

The Springer Series in Optical Sciences, under the leadership of Editor-in-Chief *William T. Rhodes*, Georgia Institute of Technology, USA, provides an expanding selection of research monographs in all major areas of optics: lasers and quantum optics, ultrafast phenomena, optical spectroscopy techniques, optoelectronics, quantum information, information optics, applied laser technology, industrial applications, and other topics of contemporary interest.

With this broad coverage of topics, the series is of use to all research scientists and engineers who need up-to-date reference books.

The editors encourage prospective authors to correspond with them in advance of submitting a manuscript. Submission of manuscripts should be made to the Editor-in-Chief or one of the Editors. See also www.springer.com/series/624

Mirosław Kozłowski
Janina Marciak-Kozłowska

Thermal Processes Using Attosecond Laser Pulses

When Time Matters

With 45 Illustrations

 Springer

Mirosław Kozłowski
Janina Marciak-Kozłowska
Institute of Electron Technology
32/46 Al. Lotników
Warsaw, Poland
miroslawkozlowski@aster.pl
kozlo@ite.waw.pl

ISBN: 978-1-4419-2136-9   e-ISBN: 978-0-387-30234-8

Printed on acid-free paper.

© 2010 Springer Science+Business Media, LLC
All rights reserved. This work may not be translated or copied in whole or in part without the written permission of the publisher (Springer Science+Business Media, LLC, 233 Spring Street, New York, NY 10013, USA), except for brief excerpts in connection with reviews or scholarly analysis. Use in connection with any form of information storage and retrieval, electronic adaptation, computer software, or by similar or dissimilar methodology now known or hereafter developed is forbidden.
The use in this publication of trade names, trademarks, service marks, and similar terms, even if they are not identified as such, is not to be taken as an expression of opinion as to whether or not they are subject to proprietary rights.

Printed in the United States of America.   (MVY)

9 8 7 6 5 4 3 2 1

springer.com

To Marcinek with love and hope

Preface

This book is based on the results of our interest in the field of ultrashort laser pulses interaction with matter. The aim of our monograph was to build the balanced description of the thermal transport phenomena generated by laser pulses shorter than the characteristic relaxation time. In the book we explore the matter on the quark, nuclear as well atomic scales. Also on the cosmic scale (Planck Era) the thermal disturbance shorter than the Planck time creates the new picture of the Universe.

The mathematics, especially PDE, are the main tool in the description of the ultrashort thermal phenomena. Two types of the PDE: parabolic and hyperbolic partial differential equations are of special interest in the study of the thermal processes.

We assume a moderate knowledge of basic Fourier and d'Alembert equations. The scope of the book is deliberately limited to the background of the quantum mechanics equations: Schrödinger and Klein-Gordon.

In this book the attosecond laser pulses are the main source of the disturbance of the thermal state of the matter. Recently, the attosecond laser pulses constitute a novel tool for probing processes taking place on the time scale of electron motion inside atoms.

The research presented in this book appears to provide the basic tools and concepts for attosecond thermal dynamics. Nevertheless much research is still needed to make this emerging field routinely applicable for a broad range of processes on atomic and subatomic scales.

Janina Marciak-Kozlowska

Miroslaw Kozlowski

Institute of Electron Technology, January 2006

Contents

Introduction

The attosecond time scale plays an important role in electronic and nuclear processes. Over the next decade, we can expect many fast electronic processes to be measured and unraveled such as Auger decay in atoms, charge-transfer reactions in molecules, and the electron dynamics of surface processes. Attosecond technology will also play a key role in resolving spatial and temporal dynamics on an atomic level with enormous potential for chemistry and the life sciences.

The contemporary achievements of attoscience are impressive. Let us summarize the most important. P. Abbamonte et al. [1] investigated the density disturbances in water with 41.3 attosecond time resolution. A. Baltuska et al. [2] present the results of the attosecond control of electronic processes by intense light fields. In 2005, A. Föhlisch et al. [3] investigated the dynamics of ultrafast electron transfer – a process important in photo- and electrochemistry.

From a theoretical point of view, the attosecond domain is the domain of application quantum mechanics. The development of attoscience is based on the theory of lasers and interaction of electromagnetic field with matter on the atomic scale.

But attoscience is the fundamental field in a broader sense than we can imagine. Let us consider that for elementary particles, e.g. electrons, pions of one second appear as long as the age of the Universe does to us.

With the attosecond "camera," the world looks and behaves quite differently. The disturbance of matter that lasts an attosecond creates a response of the matter – change of the density, temperature, elasticity – that does not follow macroscopic physics transport laws.

The measure of the response of the matter is the ratio $\delta t = \Delta t / \tau$ where Δt is the duration of the disturbance and τ is the relaxation time. For $\delta \gg 1$, all the macroscopic law, e.g., Fourier law for thermal phenomena, are valid. For $\delta \ll 1$, the new theory of transport phenomena must be developed.

This monograph is devoted to the formulation of the theoretical framework for $\delta \ll 1$ thermal phenomena. For all homogenous[1] materials we can imagine, the measured relaxation time is in the range 10^{-14} s to 10^{-9} s [4]. For all these materials, attosecond disturbance generates the $\delta \ll 1$ processes.

J.C. Maxwell [5] showed that all processes with $\delta \ll 1$ are nonstationary processes for the stationary state and can be reached after a few τ.

Later on, O. Heaviside [6] formulated the nonstationary equations for electromagnetic phenomena. In electromagnetism, the new Heaviside equation for a long time was called "telegraph equation." The name is quite misleading, for the Heaviside type equation is used in theory of evolution [7] and even stock prices [8].

As will be shown in the monograph also, the Dirac equation is the Heaviside equation for imaginary time.

In the first two chapters of the monograph, the classical theory of the thermal phenomena with $\delta \ll 1$ will be presented. In Chapter 3, the quantum Heaviside equation will be formulated and solved.

In Chapters 4, 5, and 6, the application of the attosecond Heaviside equation is presented. The interaction of the attosecond laser pulses with matter on different levels – nuclear, atomic, molecular, and cosmos – is investigated.

It occurs that contemporary laser science is developing basically with the development of a new huge laser machine, XFEL, (X-ray Free Electron Laser) [9], LUX [10], and LCLS (Liniac Coherent Light Source) [11]. The short description of these machines and LASETRON will be presented in Chapter 7.

In Chapter 8 the cosmic thermal processes with $\delta \ll 1$ will be discussed. The quantum Heaviside equation for the Planck era will be formulated and solved. In the context of the monograph, Planck era is defined as the time period $\Delta t < T_{\text{Planck}}$, where $T_{\text{Planck}} = \tau_{\text{relaxation}}$ for Planck gas.

In the scientific literature there exists a series of monographs devoted to the study the nonstationary thermal processes. The group of Prof. D. Jou published the monograph entitled *Extended Irreversible Thermodynamics* [4].

---

[1] For nonhomogenous materials, the definition of the relaxation time is not so obvious [compare W. Roetzel et al.: Int. J. Th. Sci., **42**, 541 (2003).

Within the formalism of the EIT, the Heaviside thermal equation is the master equation for heat transfer. The D.X. Tzou monograph *Macro- to Micro-Scale Heat Transfer* [13] presents the overview and the applications of the generalized Heaviside equation.

In summary, the attosecond laser pulse technology opens new possibilities in experimental as well as theoretical solid-state physics, optics, thermodynamics, and electronics. It is the first step in the new visualization of the fundamental processes that shape our Universe.

References

1. P. Abbamonte et al.: Phys. Rev. Lett. **92**, 237401 (2004)

2. A. Baltuska et al.: Nature **421**, 611 (2003)

3. A. Föhlisch et al.: Nature **436** 373 (2005)

4. D. Jou at al.: *Extended Irreversible Thermodynamics* (Springer 2001)

5. J.C. Maxwell: Philosophical Transactions London **157**, 49 (1967)

6. O. Heaviside: *Philosophical Magazine*, August 1876

7. R. Hersh: The Mathematical Intelligencier **25**, 52 (2003)

8. B. M. Bibby, M. Sorensen: Finance and Stochastics **1**, 25 1997

9. XFEL, The European X-Ray Laser Project, X-Ray Free Electron Laser, DESY (2005)

10. LUX – a Liniac/Laser Based Ultrafast X-ray Facility, http://lux.lbl.gov/Lawrence Berkley Lab

11. LCLS, Liniac Coherent Light Source, Stanford Linear Acceleration Center, http://www.ssrl.slac.stanford.edu/lcls

12. A.F. Kaplan, P.L. Shkolnikov: Phys. Rev. Lett. **88**, 074801-1 (2002)

13. D.X.Tzou: *Macro- To Micro-Scale Heat Transfer: Past, Present and Future* (Taylor and Francis USA 1996)

1

Wave Phenomena: An Overview

1.1 Hyperbolic Partial Differential Equations and Wave Phenomena

1.1.1 Equations of Wave Phenomena

One of the best models in mathematical physics is Fourier's model for heat conduction in solids. Despite the excellent agreement obtained between theory and experiment, the Fourier model contains several inconsistent implications. The most important is that the model implies an infinite speed of propagation for heat.

Despite such an unacceptable notion of energy transport in solids, the classical diffusion theory gives quite reliable results for most situations encountered in contemporary macroscopic heat transport phenomena. However, there are situations such as those dealing with extremely short time responses or very small structures where the classical diffusion model breaks down and the wave nature of heat propagation becomes dominant.

The earliest recorded speculations of the existence of propagating temperature waves was advanced by Nernst [1.1] in 1917. He suggested that in good thermal conductions at low temperature, heat may have sufficient "inertia" to give rise to an oscillating behavior.

For several of the wave phenomena that we might see or experience around us, we can derive the partial differential equations that govern their behavior. Even though the primary physical factors that give rise to these wave phenomena differ, by examing these partial differential equations it is possible to find some common form for them. Having said this, what we call *wave phenomena* come in various types, and moreover they are quite diverse. However, the reason that the characteristic of a wave is clearly common to these phenomena can be understood readily by looking at the form of the partial differential equations that govern the phenomena [1.2].

In the set of all partial differential equations, there exists a class of partial differential equations called *equations of hyperbolic type* (or simply *hyperbolic equations*). The above-mentioned equations of wave phenomena belong to this class. Now in this paragraph, we introduce several fundamental concepts of this class of equations.

As the examples of the wave equations, we consider: the equation of string vibration, the equation of oscillating of a spring, the equation for the propagation of sound, and the hyperbolic heat conduction equation.

For the vibration of string the equation, the equation of motion can be written as [1.3]:

$$\varrho(x)\frac{\partial^2 u(t,x)}{\partial t^2} = T\frac{\partial^2 u(t,x)}{\partial x^2}, \tag{1.1}$$

where $\varrho(x)$ is the density of the string at the point x, T is the tension of a string, and $u(t,x)$ is the displacement of the string.

For the oscillation of a spring we have [1.3]

$$\varrho\frac{\partial^2 u(t,x)}{\partial t^2} = k\delta^{-1}\frac{\partial^2 u(t,x)}{\partial x^2}, \tag{1.2}$$

In equation (1.2)

$$\delta = \frac{l}{b-a}, \tag{1.3}$$

l is the length of the spring, and $[a,b]$ is the length interval for the spring. Density of the spring is equal $\varrho(x)$, tension force is denoted as T, and k is the Hooke's coefficient.

Sound as the propagation of motion of air is a familiar concept. If we denote the air density $\varrho(t,x)$, then

$$\varrho(t,x) = \varrho_0(1 + u(t,x)), \tag{1.4}$$

where ϱ_0 is unperturbed air density. For $u(t,x)$, the sound propagation equation can be written as [1.3]

$$\frac{\partial^2 u(t,x)}{\partial t^2} - K\frac{\partial^2 u}{\partial x^2} = 0 \tag{1.5}$$

and $\sqrt{K}$ is the speed of propagation of the sound wave.

In the air, we can observe and measure besides the wave phenomenon the sound wave, also the diffusion – for example the diffusion of the heat. But can we observe the heat waves? For a long time the discussion of the wave, like versus diffusive behavior, was a pure mathematical problem. The standard diffusion equation doesn't take into account any propagation speed, so it cannot really be a fundamental description of the transport of heat; according to this equation if the heat source is applied to one end of the container, the temperature at the other end begins to charge instantaneously. J.C. Maxwell, working from kinetic theory, imported a ballistic term into the equation of heat conduction and ended up with hyperbolic heat conduction equation (it has first and second derivatives with respect to time), with its trade-off between the diffusive behavior (which comes from the first time derivative) and ballistic (wave) behavior (coming from the second derivative). Maxwell dropped this ballistic term after concluding that it "may be neglected, as the rate of conduction will rapidly establish itself." But as far back as the 1960s ballistic (wave) heat pulses were observed. The ballistic (wave) heat pulses are described by Maxwell-Cattaneo [1.4] equation

$$\tau\frac{\partial^2 T}{\partial t^2} + \frac{\partial T}{\partial t} = (v^2\tau)\frac{\partial^2 T}{\partial x^2}. \tag{1.6}$$

In equation (1.6), τ is the relaxation time for heat phenomena, v is the velocity of heat propagation,

$$v = \sqrt{\frac{k}{\tau \varrho c}}, \tag{1.7}$$

k is the thermal conductance, c, ϱ are the specific heat, and ϱ the density of material. Zero relaxation time for heat current, i.e. immediate reaction to change the temperature gradients, therefore leads to infinite speed as predicted by Fourier law of conduction of heat. For $\tau \to 0$, one obtains from (1.6)

$$\frac{\partial T}{\partial t} = D\frac{\partial^2 T}{\partial x^2} \tag{1.8}$$

i.e. Fourier law with diffusion coefficient

$$D = \lim_{\substack{\tau \to 0, \\ v \to \infty}} (v^2\tau). \tag{1.9}$$

As shown in Appendix A, (1.8) is the parabolic differential equation for $D > 0$. We conclude from formula (1.9) that for D finite, $v \to \infty$, i.e. $v/c \to \infty$ (c is velocity of light) in the contradiction with special relativity theory (SRT). In that case, Fourier equation can't be the fundamental equation for heat transport phenomena.

1.1.2 Hyperbolic Partial Differential Operators

Consider the following $N \times N$ matrix,

$$A_{jl}(t, x), \quad (j, l = 1, 2, \ldots, n) \tag{1.10}$$

$$H_j(t, x), \quad (j = 0, 1, \ldots, n)$$

$$A_j(t, x), \quad (j = 1, 2, \ldots, n)$$

whose entries are elements $\mathbb{R} \times \mathbb{R}^n$.

Further, suppose that $u_j(t, x)(j = 1, 2, \ldots, N)$ are unknown functions, and set $u(t, x) = (u_1(t, x), u_2(t, x), \ldots u_N(t, x))$. Finally consider the partial differential operator P given below that acts on u:

$$P(u) = \frac{\partial^2 u}{\partial t^2} + \sum_{j,l=1}^{n} A_{j,l} \frac{\partial^2 u}{\partial x_j \partial x_l} + \sum_{j=1}^{n} A_j \frac{\partial u}{\partial x_j} + \sum_{j=1}^{n} 2H_j \frac{\partial^2 u}{\partial x_j \partial t} + H_0 \frac{\partial u}{\partial t} + A_0 u.$$
$$\tag{1.11}$$

If we look back on the wave equation (1.5) then we may recover the equation from the above by setting $n = 3, N = 1$

$$A_{j,l}(t, x) = -K\delta_{jl} \tag{1.12}$$

and setting all of H_j, A_j, and A_0 to zero. In the above, δ_{jl} is the usual Kronecker delta, namely

$$\delta_{jl} = \begin{cases} 1 \ j = l, \\ 0 \ j \neq l. \end{cases} \tag{1.13}$$

For a second-order partial differential operator P, we will say that the *principal part* of P is

$$P_0 = \frac{\partial^2 I_N}{\partial t^2} + 2\sum_{j=1}^{n} H_j \frac{\partial^2}{\partial x_j \partial t} + \sum_{j,l=1}^{n} A_{j,l} \frac{\partial^2}{\partial x_j \partial x_l}. \tag{1.14}$$

For $\lambda \in \mathbb{C}$ and $\xi = (\xi_1, \xi_2, \ldots, \xi_n) \in \mathbb{R}^n$ let

$$P_0(t, x, \lambda, \xi) = \det\left(\lambda^2 I_N + 2\sum_{j=1}^{n} H_j(t, x)\xi_j\lambda + \sum_{j,l=1}^{n} A_{jl}(t, x)\xi_j\xi_l\right). \tag{1.15}$$

$P_0(t, x, \xi)$ is called the *characteristic polynomial* of the partial differential operator P. If we consider $(t, x, \xi) \in \mathbb{R} \times \mathbb{R} \times \mathbb{R}^n$ to be a parameter and P_0 to be a polynomial in λ, then the degree of the characteristic polynomial is 2N.

We denote the roots of $P_0(t, x, \lambda, \xi) = 0$ by $\lambda_k(t, x, \xi), (k = 1, 2, \ldots, 2N)$; we call these roots the *characteristic roots* of the partial differential operator P.

Now we have the definition:

The second-order partial differential operator P given in (1.11) is said to be of *hyperbolic type* (or simply just *hyperbolic*) in t-direction if for an arbitrary parameter (t, x, ξ) the characteristic roots of $P, \lambda_k(t, x, \xi)$ $(k = 1, 2, \ldots, 2N)$ are all real.

1.1.3 Formulae for Solutions of an Initial Value Problem for a Wave Equation

Let us recall the solutions to the simple classical wave equation with constant coefficient in one, two, and three dimensions [1.3, 1.4]

For one-dimensional wave phenomena one obtains:

$$\frac{1}{v^2}\frac{\partial^2 u}{\partial t^2} = \frac{\partial^2 u}{\partial x^2}, \tag{1.16}$$

where v is the signal propagation speed and where initial conditions for $u(xt)$ are set at $t = 0$ as

$$u(x, 0) = f(x), \tag{1.17}$$

$$\frac{\partial u(x, 0)}{\partial t} = g(x).$$

This has the solution of D'Alembert

$$u(x, t) = \frac{f(x + vt) + f(x - vt)}{2} + \frac{1}{2v}\int_{x-vt}^{x+vt} g(y) dy. \tag{1.18}$$

In two dimensions

$$\frac{1}{v^2}\frac{\partial^2 u}{\partial t^2} = \frac{\partial^2 u}{\partial x^2} + \frac{\partial^2 u}{\partial y^2} \tag{1.19}$$

with initial conditions at $t = 0$ for $u(x, y, t)$ of

$$u(x, y, 0) = g(x, y), \tag{1.20}$$

$$\frac{\partial u}{\partial t}(x, y, 0) = g(x, y).$$

This has the solution of Poisson

$$u(x, y, t) = \frac{1}{2\pi v}\frac{\partial}{\partial t}\iint_{\varrho \le vt}\frac{f(\xi, \eta)d\xi d\eta}{(v^2t^2 - \varrho^2)^{\frac{1}{2}}}$$

$$+ \frac{1}{2\pi c}\iint_{\varrho \le vt}\frac{g(\xi, \eta)d\xi d\eta}{(v^2t^2 - \varrho^2)^{\frac{1}{2}}}, \tag{1.21}$$

where $\varrho^2 = \left[(\xi - x)^2 + (\eta - y)^2\right]$. For three-dimensional wave phenomena

$$\frac{1}{v^2} \frac{\partial^2 u}{\partial t^2} = \frac{\partial^2 u}{\partial x^2} + \frac{\partial^2 u}{\partial y^2} + \frac{\partial^2 u}{\partial z^2} \tag{1.22}$$

with initial conditions at $t = 0$ for $u(x, y, z, t)$ of

$$u(x, y, z, 0) = f(x, y, z), \tag{1.23}$$

$$\frac{\partial u}{\partial t}(x, y, z, 0) = g(x, y, z).$$

This has the solution of Kirchhoff

$$u(x, y, z, t) = \frac{1}{4\pi v^2} \frac{\partial}{\partial t} \left(\frac{1}{t} \iint_{r=vt} f(\varepsilon, \eta, \zeta) ds \right)$$

$$+ \frac{1}{4\pi v^2 t} \iint_{r=vt} g(\xi, \eta, \zeta) d\zeta, \tag{1.24}$$

where $r^2 = (\xi - x)^2 + (\eta - y)^2 + (\zeta - z)^2$.

From these three solutions (1.18, 1.21, 1.22), something remarkable emerges. We see that in one- and two-dimensional cases, the domain of dependence that determines the solution $u(x, t)$ at point (x, t) is given by the closed interval $[x - vt, x + vt]$ and the disk (interior plus boundary) $r \leq vt$, respectively. Therefore in both cases, the signals may propagate at any speed less than or equal to v. In complete contrast, the three-dimensional solution has a domain of dependence consisting only of the *surface* of the sphere radius vt. All three-dimensional wave phenomena travel *only* at the wave velocity v. (For the moment we are ignoring the effects of dispersion here.)

What this means in practice is that in two-dimensional spaces, wave signals emitted at different times can be received simultaneously: signal reverberation occurs. It is impossible to transmit *sharply* defined signals in two dimensions.

Now let us consider one-dimensional hyperbolic partial differential equation with constant coefficients [1.4]

$$\frac{\partial^2 u(t, x)}{\partial x^2} - v^{-2} \frac{\partial^2 u(t, x)}{\partial t^2} - a \frac{\partial u(x, t)}{\partial t} - bu(t, x) = 0 \tag{1.25}$$

(i.e. one-dimensional analog of the equation $Pu = 0$, formula (1.11)).

Equation (1.25) is the Heaviside type equation. It will be shown that the term $a\frac{\partial u}{\partial t}$ represents *dissipation* of energy or damping, and the term bu represents dispersion. Setting

$$u(x, t) = f(t)g(x, t) \tag{1.26}$$

in equation (1.25) yields

$$f\frac{\partial^2 g}{\partial x^2} - v^{-2}f\frac{\partial^2 g}{\partial t^2} - \left(2v^{-2}\frac{df}{dt} + af\right)\frac{\partial g}{\partial t}$$

$$- \left(v^{-2}\frac{d^2 f}{dt^2} + a\frac{df}{dt} + bf\right)g = 0. \tag{1.27}$$

Thus $f(t)$ is arbitrary, so that we may choose it such that the coefficient of $\frac{\partial g}{\partial t}$ term vanishes. This gives $2v^{-2}\frac{df}{dt} + af = 0$, which yields

$$f(t) = e^{-\frac{av^2 t}{2}}. \tag{1.28}$$

Then the partial differential equation (PDE) has no first-derivative term and reduces to

$$\frac{\partial^2 g}{\partial x^2} - v^{-2}\frac{\partial^2 g}{\partial t^2} + kg = 0 \tag{1.29}$$

$$k = b - \frac{a^2 c^2}{4}.$$

Therefore from formula (1.26) we obtain

$$u(x,t) = e^{-\frac{av^2 t}{4}}v(x,t). \tag{1.30}$$

For $k = 0$, (1.29) becomes the wave equation for $g(x,t)$ so that

$$u(x,t) = e^{-\frac{av^2 t}{2}}\left[F(x - vt) + G(x + vt)\right]. \tag{1.31}$$

For $a > 0$, formula (1.31) represents the *damped,* undistorted wave. Suppose $k = 0$. Then if we look for a solution of the form $g = F(x - vt)$ we find that

$$\frac{d^2 F}{dt^2} - \frac{d^2 F}{dt^2} + kF = 0 \tag{1.32}$$

So that $F = 0$ if $k \neq 0$.

Similarly, $G(x + ct)$ is there. Therefore, the waves are not undistorted. However, if we set $g = F(x - \gamma t)$, where γ is different from v, we find

$$\frac{d^2 F}{dt^2} + \frac{kv^2}{v^2 - \gamma^2}F = 0. \tag{1.33}$$

This tell us that we have only sines, cosines, or hyperbolic functions as long as $v \neq \gamma$. The waves do not propagate with the constant velocity (because the parameter γ may vary). In fact, if we multiply the solution for g by some function $A(\gamma)$ and integrate over γ, this is equivalent to using the Fourier transform. Waves of this nature ($k \neq 0$) are called *dispersive waves* because they travel with variable phase velocity. The term kg in (1.29) produces *dispersion.*

1.2 The Discrete Boltzmann Equation for the Heat Transport Induced by Ultrashort Laser Pulses

1.2.1 The Model Equation

Recently it has been shown that after optical excitation by femtosecond pulse, establishment of an electron temperature by e-e scattering takes place on a few hundred femtosecond time scale in both bulk and nanostructured noble materials [1.5]–[1.8]. In noble metal clusters, the electron thermalization time (relaxation time) is of the order 200 fs [1.7, 1.8]. This relaxation time is much larger than the duration of the now available femtosecond optical pulses offering the unique possibility of analyzing the properties of a thermal quasi-free electron gas [1.9]. In paper [1.9] using a two-color femtosecond pump-probe laser technique, the ultrafast energy exchanges of a nonequilibrium electron gas was investigated. When the duration of the laser pulse, 25 fs, in paper [1.9], is shorter than the relaxation time, the parabolic Fourier equation cannot be used [1.10, 1.11]. Instead, the new hyperbolic quantum heat transport equation is the valid equation [1.12]. The quantum heat transport equation is the wave damped equation for heat phenomena on the femtosecond scale.

Wave is an organized propagating imbalance [1.14]. Some phenomena seem to be clearly diffusive, with no wave-like implications, heat for instance. That was consistent with experiments in the past century, but not any longer. As far back as the 1960s, ballistic (wave-like) heat pulses were observed at low temperatures [1.14]. The idea was that heat is just the manifestation of microscopic motion. Computing the classical resonant frequencies of atoms or molecules in a lattice gives numbers of the order 10^{13} Hz, which is in the infrared, so when molecules jiggle they give off heat. These lattice vibrations are called phonons. Phonons have both wave-like and particle-like aspects. Lattice vibrations are responsible for the transport of heat, and we know that is a diffusive phenomenon, described by the Fourier equation. However, if the lattice is cooled to near absolute zero, the mean free scattering of the phonons becomes comparable to the macroscopic size of the sample. When this happens, lattice vibrations no longer behave diffusively but are actually wave-like or thermal wave. By controlling the temperature of a sample, one can control the extent to which heat is ballistic (thermal wave) or diffusive. In essence, if a heat pulse is launched into sample (by the laser pulse interaction) and if the phonons can get across the sample without scattering, they will propagate as thermal waves.

The extent to which the motion of quasiparticles (phonons) or particles is ballistic is described by the value of the relaxation time, τ. For ballistic (wave-like) motion, $\tau \to \infty$. The equation that is a generalization of the Fourier equation (in which $\tau \to \infty$) is the Heaviside equation [1.10] for thermal processes:[1]

$$\tau \frac{\partial^2 T}{\partial t^2} + \frac{\partial T}{\partial t} = D\nabla^2 T. \tag{1.34}$$

For very short relaxation time $\tau \to 0$ we obtain from (1.34) the Fourier equation

$$\frac{\partial T}{\partial t} = D\nabla^2 T \tag{1.35}$$

and for $\tau \to \infty$ we obtain from formula (1.35), the ballistic $\equiv$ thermal wave motion:

$$\frac{1}{v^2} \frac{\partial^2 T}{\partial t^2} = \nabla^2 T. \tag{1.36}$$

In the set of papers [1.10, 1.11, 1.12], the quantum generalization of the Heaviside equation was obtained and solved:

$$\frac{1}{v^2} \frac{\partial^2 T}{\partial t^2} + \frac{m}{\hbar} \frac{\partial T}{\partial t} = \frac{\partial^2 T}{\partial t^2}, \tag{1.37}$$

where $v = \alpha c$ and is the fine structure constant, and c is the vacuum light velocity. In formula (1.37), m is the *heaton* mass [1.7]. *Heaton* energy is equal E_h

$$E_h = m\alpha^2 c^2. \tag{1.38}$$

In papers [1.10], the Heaviside equation was obtained for the fermionic gases (electrons, nucleons, quarks). In this paragraph, the Heaviside equation will be obtained for particles with mass m, where m is the mass of the fermion or boson. Moreover, beside the elastic scattering of the particles, the creation and absorption of the heat carriers will be discussed. The new form of the discrete Heaviside equation can be obtained as the result of the discretization of the one-dimensional Boltzmann equation. The solution of the discrete Boltzmann equation will be obtained for Cauchy boundary conditions, initialed by ultrashort laser pulses, i.e. for $\Delta t \leq \tau$, the relaxation time.

---

[1] For the first time, linear hyperbolic equation of the type $\frac{\partial^2 v}{\partial x^2} = kc\frac{\partial v}{\partial t} + sc\frac{\partial^2 v}{\partial t^2}$ was formulated by Oliver Heaviside, see: O. Heaviside, *Philosophical Magazine*, 1876 and O. Heaviside: *Electrical Papers*, Chelsea Publishing Company, New York 1970, pp 53–61.

Let us consider the one-dimensional rod (wire) that can transport "particles" – heat carriers [1.15]. These particles, however, may move only to the right or to the left on the rod. Moving particles may collide with the fixed (scatter centra, barriers, dislocations) the probabilities of such collisions and their expected results being specified. All particles will be of the same kind, with the same energy and other physical specifications distinguishable only by their direction.

Let us define:

$u(z,t)$ = expected density of particles at z and at time t moving to the right,

$v(z,t)$ = expected density of particles at z and at time t moving to the left.

Furthermore, let $\sigma(z)$ = probability of collision occurring between a fixed scattering centrum and a particle moving between z and $z + \Delta$.

Suppose that a collision might result in the disappearance of the moving particle without new particle appearing. Such a phenomenon is called *absorption*. Or the moving particle may be reversed in direction or back-scattered. We shall agreeing that in each collision at z, an expected total of $F(z)$ particles arises moving in the direction of the original particle, and $B(z)$ arise going in the opposite direction.

The expected total number of right-moving particles in $z_1 \leq z \leq z_2$ at time t is

$$\int_{z_1}^{z_2} u(z,t)\mathrm{d}z\,, \qquad (1.39)$$

while the total number of particles passing z to the right in the time interval $t_1 \leq t \leq t_2$ is

$$w \int_{t_1}^{t_2} u(z,t)\mathrm{d}t \qquad (1.40)$$

where w is the particles speed.

Consider the particle moving to the right and passing $z + \Delta$ in the time interval $t_1 + \frac{\Delta}{w} \leq t \leq t_2 + \frac{\Delta}{w}$:

$$w \int_{t_1+\Delta/w}^{t_2+\Delta/w} u(z+\Delta,t')\mathrm{d}t' = w \int_{t_1}^{t_2} u\left(z+\Delta, t' + \frac{\Delta}{w}\right)\mathrm{d}t'. \qquad (1.41)$$

These can arise from particles that passed z in the time interval $t_1 \leq t \leq t_2$ and came through $(z, z+\Delta)$ without collision

$$w \int_{t_1}^{t_2} (1 - \Delta\sigma(z,t'))u(z,t')\mathrm{d}t' \qquad (1.42)$$

plus contributions from collisions in the interval $(z, z+\Delta)$. The right-moving particles interacting in $(z, z+\Delta)$ produce in the time t_1 to t_2,

$$w \int_{t_1}^{t_2} \Delta\sigma(z,t')F(z,t')u(z,t')dt' \tag{1.43}$$

particles to the right, while the left moving ones give:

$$w \int_{t_1}^{t_2} \Delta\sigma(z,t')B(z,t')v(z,t')dt'. \tag{1.44}$$

Thus

$$w \int_{t_1}^{t_2} u\left(z + \Delta, t' + \frac{\Delta}{w}\right) dt' = w \int_{t_1}^{t_2} u(z,t')dt'$$

$$+ w\Delta \int_{t_1}^{t_2} \sigma(z,t')(F(z,t') - 1)u(z,t')dt'$$

$$= +w\Delta \int_{t_1}^{t_2} \sigma(z,t')B(z,t')v(z,t')dt'. \tag{1.45}$$

Now, we can write:

$$u\left(z + \Delta, t' + \frac{\Delta}{w}\right) = u(z,t') + \left(\frac{\partial u}{\partial z}(z,t') + \frac{1}{w}\frac{\partial u}{\partial t}(z,t')\right)\Delta \tag{1.46}$$

to get

$$\int_{t_1}^{t_2} \left(\frac{\partial u}{\partial z}(z,t')\right) + \frac{1}{w}\frac{\partial u}{\partial t}(z,t')\,dt' = \int_{t_1}^{t_2} \sigma(z,t')((F(z,t') - 1)u(z,t')$$

$$+ B(z,t')v(z,t'))dt'. \tag{1.47}$$

On letting $\Delta \to 0$ and differentiating with respect to t_2, we find

$$\frac{\partial u}{\partial z} + \frac{1}{w}\frac{\partial u}{\partial t} = \sigma(z,t)(F(z,t) - 1)u(z,t)$$

$$+ \sigma(z,t)B(z,t)v(z,t). \tag{1.48}$$

In a like manner

$$-\frac{\partial v}{\partial z} + \frac{1}{w}\frac{\partial v}{\partial t} = \sigma(z,t)B(z,t)u(z,t)$$

$$+ \sigma(z,t)(F(z,t) - 1)v(z,t). \tag{1.49}$$

The system of partial differential equations of hyperbolic type (1.48, 1.49) is the Boltzmann equation for one-dimensional transport phenomena [1.15].

Let us define the total density for heat carriers, $\varrho(z,t)$

$$\varrho(z,t) = u(z,t) + v(z,t) \tag{1.50}$$

and density of heat current

$$j(z,t) = w(u(z,t) - v(z,t)).$$ (1.51)

Considering (1.48–1.51) one obtains

$$\frac{\partial \varrho}{\partial z} + \frac{1}{w^2}\frac{\partial j}{\partial t} = \sigma(z,t)u(z,t)(F(z,t) - B(z,t) - 1)$$

$$+ \sigma(z,t)v(z,t)(B(z,t) - F(z,t) + 1).$$ (1.52)

Equation (1.52) can be written as

$$\frac{\partial \varrho}{\partial z} + \frac{1}{w^2}\frac{\partial j}{\partial t} = \frac{\sigma(z,t)(F(z,t) - B(z,t) - 1)j}{w}$$ (1.53)

or

$$j = \frac{w}{\sigma(z,t)(F(z,t) - B(z,t) - 1)}\frac{\partial \varrho}{\partial z}$$

$$+ \frac{1}{w\sigma(z,t)(F(z,t) - B(z,t) - 1)}\frac{\partial j}{\partial t}.$$ (1.54)

Denoting D, diffusion coefficient

$$D = -\frac{w}{\sigma(z,t)(F(z,t) - B(z,t) - 1)}$$

and τ, relaxation time

$$\tau = \frac{1}{w\sigma(z,t)(1 - F(z,t) - B(z,t))}$$ (1.55)

equation (1.54) takes the form

$$j = -D\frac{\partial \varrho}{\partial z} - \tau\frac{\partial j}{\partial t}.$$ (1.56)

Equation (1.56) is the Cattaneo type equation and is the generalization of the Fourier equation. Now in a like manner we obtain from (1.48–1.51)

$$\frac{1}{w}\frac{\partial j}{\partial z} + \frac{1}{w}\frac{\partial \varrho}{\partial t} = \sigma(z,t)u(z,t)(F(z,t) - 1 + B(z,t))$$

$$+ \sigma(z,t)v(z,t)(B(z,t) + F(z,t) - 1))$$ (1.57)

or

$$\frac{\partial j}{\partial z} + \frac{\partial \varrho}{\partial t} = 0.$$ (1.58)

Equation (1.58) describes the conservation of energy in the transport processes.

Considering (1.56) and (1.58) for the constant D and τ, the hyperbolic Heaviside equation is obtained:

$$\tau\frac{\partial^2 \varrho}{\partial t^2} + \frac{\partial \varrho}{\partial t} = D\frac{\partial^2 \varrho}{\partial z^2}. \tag{1.59}$$

In the case of the *heaton* gas with temperature $T(z,t)$, (1.59) has the form

$$\tau\frac{\partial^2 T}{\partial t^2} + \frac{\partial T}{\partial t} = D\frac{\partial^2 T}{\partial z^2},$$

where τ is the relaxation time for the thermal processes.

1.2.2 The Solution of the Boltzmann Equation for the Stationary Transport Phenomena in One-Dimensional Wire

In the stationary state, transport phenomena $dF(z,t)/dt = dB(z,t)dt = 0$ and $d\sigma(z,t)/dt = 0$. In that case, we denote $F(z,t) = F(z) = B(z,t) = B(z) = k(z)$ and (1.43) and (1.44) can be written as

$$\frac{du}{dz} = \sigma(z)(k-1)u(z) + \sigma(z)kv(z),$$

$$-\frac{dv}{dz} = \sigma(z)k(z)u(z) + \sigma(z)(k(z)-1)v(z) \tag{1.60}$$

with diffusion coefficient

$$D = \frac{w}{\sigma(z)} \tag{1.61}$$

and relaxation time

$$\tau(z) = \frac{1}{w\sigma(z)(1-2k(z))}. \tag{1.62}$$

The system of (1.60) can be written as

$$\frac{d^2u}{dz^2} - \frac{\frac{d}{dz}(\sigma k)}{\sigma k}\frac{du}{dz}$$

$$+ u\left[\sigma^2(2k-1) + \frac{d\sigma}{dz}(1-k) + \frac{\sigma(k-1)}{\sigma k}\frac{d(\sigma k)}{dz}\right] = 0,$$

$$\tag{1.63}$$

$$\frac{du}{dz} = \sigma(k-1)u + \sigma kv(z). \tag{1.64}$$

Equation (1.63) after differentiation has the form

$$\frac{d^2u}{dz^2} + f(z)\frac{du}{dz} + g(z)u(z) = 0 \tag{1.65}$$

where

$$f(z) = -\frac{1}{\sigma}\left(\frac{\sigma}{k}\frac{dk}{dz} + \frac{d\sigma}{dz}\right),$$

$$g(z) = \sigma^2(z)(2k-1) - \frac{\sigma}{k}\frac{dk}{dz}. \tag{1.66}$$

For the constant absorption rate we put

$$k(z) = k = \text{constant} \neq \frac{1}{2}.$$

In that case

$$f(z) = -\frac{1}{\sigma}\frac{d\sigma}{dz},$$

$$g(z) = \sigma^2(z)(zk-1). \tag{1.67}$$

With functions $f(z)$ and $g(z)$, the general solution of the (1.63) has the form

$$u(z) = C_1 e^{(1-2k)^{1/2}\int \sigma dz} + C_2 e^{-(1-2k)^{1/2}\int \sigma dz}. \tag{1.68}$$

In the subsequent, we will consider the solution of (1.65) with $f(z)$ and $g(z)$ described by (1.67) for Cauchy condition:

$$u(0) = q, \quad v(a) = 0. \tag{1.69}$$

Boundary condition (1.69) describes the generation of the heat carriers (by illuminating the left end of the strand with laser pulses) with velocity q heat carrier per second. The solution has the form:

$$u(z) = \frac{2q e^{[f(0)-f(a)]}}{1 + \beta e^{2[f(0)-f(a)]}}\left[\frac{(1-2k)^{\frac{1}{2}}}{(1-2k)^{\frac{1}{2}} - (k-1)}\right]\cosh\left[f(x) - f(a)\right],$$

$$+ \frac{k-1}{(1-2k)^{\frac{1}{2}} - (k-1)}\sinh\left[f(x) - f(a)\right]$$

$$u(z) = \frac{2q e^{(f(0)-f(a))}}{1 + \beta e^{2[f(0)-f(a)]}}\left[\frac{(1-2k)^{\frac{1}{2}} + (k-1)}{k}\sinh\left[f(x) - f(a)\right]\right],$$

$$\tag{1.70}$$

where

$$f(z) = (1-2k)^{\frac{1}{2}}\int \sigma dz,$$

$$f(0) = (1-2k)^{\frac{1}{2}}\left[\int \sigma dz\right]_0,$$

$$f(a) = (1 - 2k)^{\frac{1}{2}} \left[\int \sigma dz \right]_a,$$

$$\beta = \frac{(1 - 2k)^{\frac{1}{2}} + (k - 1)}{(1 - 2k)^{\frac{1}{2}} - (k - 1)}. \tag{1.71}$$

Considering formulae (1.50), (1.51), and (1.70), we obtain for the density $\varrho(z)$ and current density $j(z)$.

$$j(z) = \frac{2qwe^{[f(0)-f(a)]}}{1 + \beta e^{2[f(0)-f(a)]}} \left[\frac{(1 - 2k)^{\frac{1}{2}}}{(1 - 2k)^{\frac{1}{2}} - (k - 1)} \cosh\left[f(z) - f(a)\right] \right.$$

$$\left. - \frac{1 - 2k}{(1 - 2k)^{\frac{1}{2}} - (k - 1)} \sinh\left[f(z) - f(a)\right] \right] \tag{1.72}$$

and

$$q = \frac{2qe^{[f(0)-f(a)]}}{1 + \beta e^{2[f(0)-f(a)]}} \left[\frac{(1 - 2k)^{\frac{1}{2}}}{(1 - 2k)^{\frac{1}{2}} - (k - 1)} \cosh\left[f(z) - f(a)\right] \right.$$

$$\left. - \frac{1}{(1 - 2k)^{\frac{1}{2}} - (k - 1)} \sinh\left[f(z) - f(a)\right] \right]. \tag{1.73}$$

Equations (1.72) and (1.73) fulfill the generalized Fourier relation

$$j = -\frac{w}{\sigma(z)} \frac{\partial \varrho}{\partial z}, \qquad D = \frac{W}{\sigma(z)}, \tag{1.74}$$

where D denotes the diffusion coefficient.

Analogously, we define the generalized diffusion velocity $v_D(z)$

$$v_D(z) = \frac{j(z)}{n(z)} \tag{1.75}$$

$$= \frac{w(1 - 2k)^{\frac{1}{2}} \left[\cosh[f(z) - f(a)] - (1 - 2k)^{\frac{1}{2}} \sinh[f(x) - f(a)] \right]}{(1 - 2k)^{\frac{1}{2}} \cosh[f(x) - f(a)] - \sinh[f(x) - f(a)]}.$$

Assuming constant cross section for heat carriers scattering $\sigma(z) = \sigma_o$, we obtain from formula (1.71)

$$f(z) = (1 - 2k)^{\frac{1}{2}} z,$$

$$f(0) = 0,$$

$$f(a) = (1 - 2k)^{\frac{1}{2}} a \tag{1.76}$$

and for density $\varrho(z)$ and current density $j(z)$

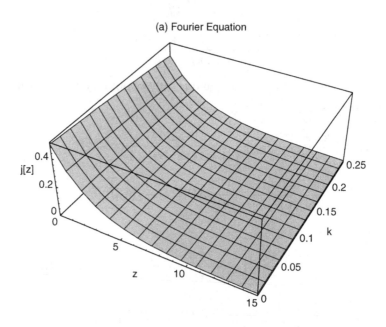

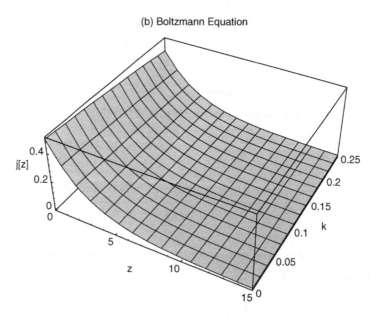

Fig. 1.1. (a) The solution of the Fourier equation for the Cauchy boundary condition (1.61). (b) The solution of the Boltzmann equation for $a \gg L$.

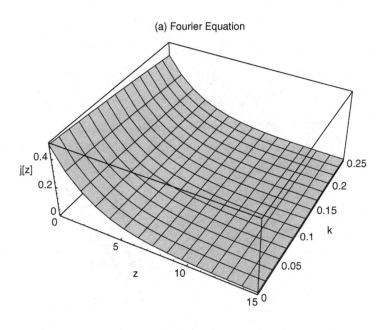

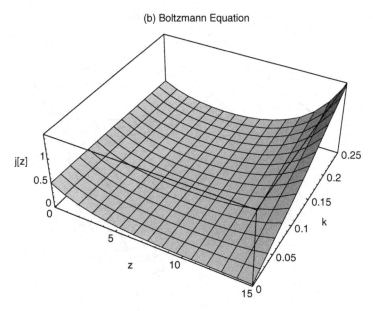

Fig. 1.2. (**a**) The same as in Fig. 1.1(a). (**b**) The solution of the Boltzmann equation for $a = L$.

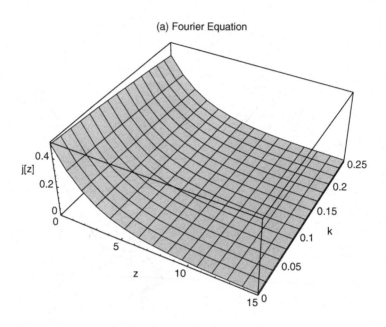

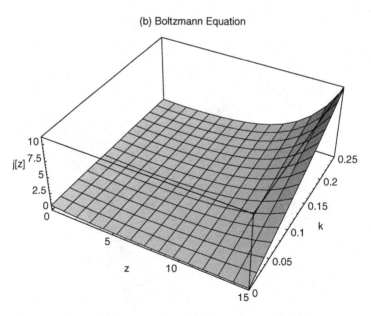

Fig. 1.3. (a) The same as in Fig. 1.1(a). (b) The solution of the Boltzmann equation for $a \ll L$.

$$j(z) = \frac{2qwe^{-(1-2k)^{\frac{1}{2}}a\sigma}}{1 + \beta e^{-(1-2k)^{\frac{1}{2}}a\sigma}} \left[\frac{(1-2k)^{\frac{1}{2}}}{(1-2k)^{\frac{1}{2}} - (k-1)} \cosh\left[(2k-1)^{\frac{1}{2}}(x-a)\sigma\right] \right.$$

$$= \left. -\frac{(1-2k)}{(1-2k)^{\frac{1}{2}} - (k-1)} \sinh\left[(2k-1)^{\frac{1}{2}}(x-a)\sigma\right] \right], \tag{1.77}$$

$$\varrho(z) = \frac{2qe^{-(1-2k)^{\frac{1}{2}}a\sigma}}{1 + \beta e^{-(1-2k)^{\frac{1}{2}}a\sigma}} \left[\frac{(1-2k)^{\frac{1}{2}}}{(1-2k)^{\frac{1}{2}} - (k-1)} \cosh\left[(2k-1)^{\frac{1}{2}\sigma}(x-a)\right] \right.$$

$$= \left. -\frac{1}{(1-2k)^{\frac{1}{2}} - (k-1)} \sinh\left[(2k-1)^{\frac{1}{2}}(x-a)\sigma\right] \right]. \tag{1.78}$$

We define Fourier's diffusion velocity $v_F(z)$ and diffusion length, L

$$v_F = \left(\frac{D}{\tau}\right)^{\frac{1}{2}}, \qquad L = v_F \tau. \tag{1.79}$$

Considering formulae (1.61) and (1.62), one obtains

$$v_F(z) = w(1-2k)^{\frac{1}{2}},$$

$$L = \frac{1}{\sigma(1-2k)^{\frac{1}{2}}} = \frac{\lambda_{mfp}}{(1-2k)^{\frac{1}{2}}} \tag{1.80}$$

where λ_{mfp} denotes the mean free path for heat carriers.

Considering formulae (1.77), (1.78), (1.79), and (1.80), one obtains

$$j(z) = \frac{2qwe^{-\frac{a}{L}}}{1 + \beta e^{-\frac{a}{L}}} \left[\frac{(1-2k)^{\frac{1}{2}}}{(1-2k)^{\frac{1}{2}} - (k-1)} \cosh\left[\frac{(x-a)}{L}\right] \right.$$

$$\left. -\frac{(1-2k)}{(1-2k)^{\frac{1}{2}} - (k-1)} \sinh\left[\frac{x-a}{L}\right] \right], \tag{1.81}$$

$$\varrho(z) = \frac{2qe^{-\frac{a}{L}}}{1 + \beta e^{-\frac{a}{L}}} \left[\frac{(1-2k)^{\frac{1}{2}}}{(1-2k)^{\frac{1}{2}} - 1} \cosh\left[\frac{x-a}{L}\right] \right.$$

$$\left. -\frac{1}{(1-2k)^{\frac{1}{2}} - (k-1)} \sinh\left[\frac{x-a}{L}\right] \right]. \tag{1.82}$$

In Figs. 1.1–1.3, we present the results of the calculation for density currents of heat carriers. Figures 1.1(a)–1.3(a) are the solution of the Fourier equation for the boundary condition (1.69). Figures 1.1(b)–1.3(b) represent the solution of the Boltzmann equation formula (1.81) for $a \gg L$, $a = L$, and $a \ll L$ respectively, when $k \in [0, 0.5]$. For the length of strand $a \gg L$, solutions of

Fourier and Boltzmann equations overlap. For $a \leq L$, the Boltzmann equation gives the different description of the transport processes. In that case, the solution of the Boltzmann equation depends strongly on the scatterings (k coefficient) of the carriers.

Recently [1.16], the heat conduction was one-dimensional system is actively investigated. As was discussed in paper [1.16], the dependence of density current on L can be described by the general formula

$$j \sim L^{\alpha},$$

where α can be positive or negative. The same conclusion can be drawn from the calculation presented in our paper. In this calculation, coefficient α depends on the scattering cross section for the heat carriers.

1.3 The Existence of a Solution for a Hyperbolic Equation and Its Properties

1.3.1 Finite Propagation Speed, Domains of Dependence and Influence

In this paragraph, we study the basic properties of a single linear second-order hyperbolic equation. By applying, for instance, an external force to some part of whatever we are considering, we can cause a disturbance. With time, this disturbance is passed onto the surrounding parts. One particular characteristic of wave phenomena is that the speed with which the disturbance propagates itself is finite. We shall show that finiteness of the speed of propagation is a property of a second-order hyperbolic equation. We will then show that a solution for the initial boundary value problem does exist. In this paragraph, we consider the following second-order hyperbolic operator:

$$P = \frac{\partial^2}{\partial t^2} + 2 \sum_{j=1}^{n} h_j(t,x) \frac{\partial^2}{\partial t \partial x_j} - \sum_{j,l=1}^{n} a_{j,l}(t,x) \frac{\partial^2}{\partial x_i \partial x_l}$$

$$+ \sum_{j=1}^{n} a_j(t,x) \frac{\partial}{\partial x_j} + h_o(t,x) \frac{\partial}{\partial t} + a_0(t,x). \tag{1.83}$$

Equation (1.83) is equation (1.15) with $N = 1$. We introduce the following notation:

$$A = \sum_{j,l=1}^{n} a_{j,l}(t,x) \frac{\partial^2}{\partial x_j \partial x_l} - \sum_{j=1}^{n} a_j(t,x) - a_0(t,x) \tag{1.84}$$

and

$$H = 2 \sum_{j=1}^{n} h_j(t,x) \frac{\partial}{\partial x_j} + h_0. \tag{1.85}$$

In terms of (1.84) and (1.85), we can write the operator P as

$$P_u = \frac{\partial^2 u}{\partial t^2} + H \frac{\partial u}{\partial t} - Au. \tag{1.86}$$

We consider the following initial boundary value problem

$$P_u = f(t,x),$$

$$B_u = g(t,x),$$

$$u(0,x) = u_0(x), \tag{1.87}$$

$$\frac{\partial u(0,x)}{\partial t} = u_1(x).$$

More explicitly, at initial time $t = 0$ the initial states u_0 and u_1 have been assigned, and we are given f and the condition g on the boundary. The problem is to find a solution that fulfills these conditions. In [1.4], the following theorem is proved:

For the initial boundary value problem (1.87), the domain of influence of

$$(t_0, x_0) C\{(t,x); |x - x_0| \leq v_{max}(t - t_0)\}. \tag{1.88}$$

Theorem (1.88) shows that for the phenomena governed by (1.87) the disturbance, at some point, propagates with time to its surroundings and with finite speed that never exceeds v_{max}. Due to the above property of (1.87), that is to say, every disturbance is transmitted with finite speed, the initial boundary value problem (1.87) is often said to have finite speed of propagation.

1.3.2 An a Priori Estimate of the Solution

Now we focus our attention on the existence of the solution of the initial boundary value problem (1.87). In this paragraph, we assume that the coefficients of the operator P and the boundary operator B do not depend on the time variable t. We should recognize that as was shown in [1.3] that even if

Table 1.1. Waves Versus Diffusion

	Property	Waves	Diffusions
(i)	Speed of propagation	Finite ($\leq c$)	Infinite
(ii)	Well posed for $t > 0$	Yes	Yes
(iii)	Well posed for $t < 0$	Yes	No
(iv)	Information	Transported	Lost gradually

they do not depend on t, the same conclusion can be derived. First of all, we describe well-posed problems of PDE, Partial Differential Equations.

Well-posed problems consist of a PDE in a domain together with a set of initial and/or boundary conditions (or other auxiliary conditions) that enjoy the following fundamental properties:

Existence: There exists at least one solution $u(x,t)$ satisfying all these conditions.

Uniqueness: There is at most one solution.

Stability: The unique solution $u(x,t)$ depends in a stable manner on the date of the problem. This means that if the data are changed a little, the corresponding solution changes only a little.

We have seen that the basic property of waves is that information gets transported in both directions at a finite speed. On the other hand, the basic property of the second transport process-diffusion is that the initial disturbance gets spread out in a smooth fashion and gradually disappears. The fundamental properties of those two equations are summarized in Table 1.1 [1.3].

For the wave equation, we have seen most of these properties already. Generally speaking, the wave equation just moves information along the characteristic lines.

As for property (i), for the diffusion equation

$$\frac{\partial u}{\partial t} = k \frac{\partial^2 u}{\partial x^2} \, , \tag{1.89}$$

we have the solution [1.3]

$$u(x,t) = \frac{1}{\sqrt{4\pi kt}} \int_{-\infty}^{\infty} e^{-\frac{(x-y)^2}{4kt}} \Phi(y) dy \, . \tag{1.90}$$

From formula (1.90), we conclude that the value of $u(x,t)$ depends on the values of the initial datum $\Phi(y)$ for *all* y where $-\infty < y < \infty$. Conversely, the value of Φ at a point x_0 has an immediate effect everywhere (for $t > 0$) even though most of its effect is only for a short time near x_0. Therefore, the speed of propagation is infinite. This is in stark contrast to the wave equation (and all hyperbolic equations).

As for (iv), there are several ways to see that the *diffusion* equation is not well posed for $t < 0$ ("backward in time"). Let

$$u_n(x,t) = \frac{1}{n}\sin(nx)e^{-n^2kt}. \tag{1.91}$$

We can check that this satisfies the diffusion equation for all x, t. Also $u_n(x,0) = n^{-1}\sin(nx) \to 0$ uniformly as $n \to \infty$. But consider any $t < 0$, say $t = -1$. Then $u_n(x,-1) = n^{-1}\sin(nx)e^{kn^2} \to \pm\infty$ uniformly as $n \to \infty$ except for a few x. Thus u_n is close to the zero solution at time $t = 0$ but not at time $t = -1$. This violates the stability in the uniform sense at last.

1.3.3 The Wave Solution of the Heat Conduction Equation

For a one-dimensional case, the classical heat conduction equation is of the form [1.3]

$$\gamma\frac{\partial T}{\partial t} = \frac{\partial}{\partial X}\left(k\frac{\partial T}{\partial X}\right). \tag{1.92}$$

If γ and k are constant, thermal conductivity may be included into a spatial coordinate with the aid of the transform $x = \sqrt{\gamma/k}X$

$$\frac{\partial T}{\partial t} = \frac{\partial^2 T}{\partial X^2}. \tag{1.93}$$

Equation (1.93) can be satisfied by two wave solutions. A rightward traveling wave is defined by the formula

$$T_1(x,t) = A\exp[-v(x - vt)] \tag{1.94}$$

and the leftward

$$T_2(x,t) = A\exp[-v(x + vt)]. \tag{1.95}$$

If further we pass over to the imaginary number $v \to in$, $A \to 2iA_n$ then the difference $T = T_2 - T_1$ will be expressed by the following formula

$$T(x,t) = A_n\exp[-n^2t]\sin nx. \tag{1.96}$$

At $t_0 \le t \le 0$, $0 \le x \le \pi$, formula (1.96) may be treated as a solution of the temperature history problem of a heated body from its state at the given moment.

It is easy to show that at $t_0 \le t \le 0$, the problem under consideration is incorrect according to Hadamard [1.17]. Let us evaluate a norm of close initial data

$$\| A_n \sin nx - 0 \| = \max_{0 \le x \le \pi} |A_n \sin nx| \le A_n. \tag{1.97}$$

Following Hadamard, let A_n be equal to $e^{-\sqrt{n}}$. For $n \to \infty$, formula (1.97) is going to zero also. Next consider a norm of solution difference satisfying the initial data

$$\| \exp[-(\sqrt{n}) - n^2 t] \sin nx - 0 \|$$

$$= \max_{\substack{0 \le x \le \pi \\ t_0 \le t \le 0}} | \exp[-(\sqrt{n}) - n^2 t] \sin nx| \le \frac{\exp[n^2(t_0)]}{\exp[\sqrt{n}]}.$$

Hence it follows that at $n \to \infty$ the norm of solution differences becomes infinitely large at finite t_0 when the difference between the initial data is going to zero. Therefore, the proposed problem appears to be incorrect according to Hadamard.

Let us determine a line of equal temperatures $T(x,t) = \text{const}$. The following equality is valid

$$\frac{\partial T}{\partial t} + \frac{\partial T}{\partial X}\frac{dX}{dt} = 0. \tag{1.98}$$

Designate the velocity

$$C(X,t) = \frac{dX}{dt}.$$

In the one-dimensional case, C coincides with the velocity of the isotherm, which, in its turn, is determined as a ratio between infinitesimal increment of the normal external to an isotherm and infinitesimal time interval.

It follows from (1.98) that

$$C(X,t) = -\frac{\dfrac{\partial T}{\partial t}}{\dfrac{\partial T}{\partial X}}. \tag{1.99}$$

Now from solution of (1.96), formula (1.99) takes the form

$$C = \sqrt{\frac{k}{\gamma}}\, n \tan\left[nX\sqrt{\frac{\gamma}{k}}\right] = \sqrt{\frac{k}{\gamma}}\, n \tan nx. \tag{1.100}$$

An analysis of formula (1.100) reveals that there are points where the velocity $C(X, t)$ vanishes or become infinite.

The marked peculiarities should make investigators (both experimentalists and theoreticians) be careful when employing the heat conduction equation.

The remedy for the parabolic heat conduction equation (1.92) can be found in the hyperbolic heat conduction equation. To that aim, rewrite (1.98) as

$$\frac{\partial T}{\partial t} + C(X, t) \frac{\partial T}{\partial X} . \tag{1.101}$$

Differentation of (1.101) with respect to X and t followed the elimination of the mixed derivative $\partial^2 T / \partial X \partial t$.

$$\frac{\partial^2 T}{\partial t^2} + \left(\frac{\partial v}{\partial t} - C \frac{\partial c}{\partial X} \right) \frac{\partial T}{\partial X} = C^2 \frac{\partial^2 T}{\partial X^2} . \tag{1.102}$$

By using (1.99), we eliminate $\partial T / \partial X$ and rewrite the above equality as

$$\frac{1}{v^2} \frac{\partial^2 T}{\partial t^2} + \frac{1}{v^2} \left[\frac{\partial v}{\partial X} - \frac{1}{v} \frac{\partial v}{\partial t} \right] \frac{\partial T}{\partial t} = \frac{\partial^2 T}{\partial X^2} . \tag{1.103}$$

Equation (1.103) turns into (1.92) if $v \to \infty$ and

$$\frac{1}{v^2} \left[\frac{\partial v}{\partial X} - \frac{1}{v} \frac{\partial v}{\partial t} \right] \to \frac{\gamma}{k} .$$

In case of the constant isotherm velocity, equation (1.103) is simplified to the form of the classical wave equation

$$\frac{\partial^2 T}{\partial t^2} = C^2 \frac{\partial^2 T}{\partial X^2} . \tag{1.104}$$

Let $v = v(X)$, then (1.103) acquires the form

$$\frac{1}{v^2} \frac{\partial^2 T}{\partial t^2} + \frac{1}{v^2} \frac{dv}{dX} \frac{\partial T}{\partial t} = \frac{\partial^2 T}{\partial X^2} . \tag{1.105}$$

If $(1/v^2)(dv/dX) = \gamma/k$ where γ and k are heat capacity and thermal conductivity, respectively, then

$$\left(\frac{\gamma}{k} X + \text{const} \right)^2 \frac{\partial^2 T}{\partial t^2} + \frac{\gamma}{k} \frac{\partial T}{\partial t} = \frac{\partial^2 T}{\partial X^2} .$$

Dividing both sides by γ/k, one obtains

$$\left[\sqrt{\frac{\gamma}{k}} X + \text{const} \sqrt{\frac{k}{\gamma}} \right]^2 \frac{\partial^2 T}{\partial t^2} + \frac{\partial T}{\partial t} = \frac{k}{\gamma} \frac{\partial^2 T}{\partial X^2} .$$

Because k/γ is the constant value, thermal conductivity may be induced into a spatial coordinate, with the aid of the particular transform $x = \sqrt{\gamma/k}X$; besides, in some case the coefficient at $\partial^2 T/\partial t^2$ may be approximately assumed equal to the constant parameter λ. Then finally

$$\lambda\frac{\partial^2 T}{\partial T^2} + \frac{\partial T}{\partial t} = \frac{\partial^2 T}{\partial x^2}\,. \tag{1.106}$$

By using (1.106), it is easy to obtain solutions for direct and inverse waves in the form

$$T_1(x,t) = A\exp\left[-\frac{v}{1-\lambda v^2}(x-vt)\right]\,, \tag{1.107}$$

$$T_2(x,t) = A\exp\left[\frac{v}{1-\lambda v^2}(x+vt)\right]\,.$$

Assuming $v = in$, $a = 2iA_n$, $T_2 - T_1$ yields

$$T(x,t) = T_2 - T_1 = A_n\exp\left[-\frac{n^2 t}{1+\lambda n^2}\right]\sin\frac{nx}{1+\lambda n^2}\,. \tag{1.108}$$

If is evident that for $\lambda \to 0$, solution (1.108) converges to (1.95). Next we demonstrate that close solutions correspond to close initial data

$$T_1 = \exp[-\sqrt{n}]\sin\frac{nx}{1+\lambda n^2}$$

and

$$T_2 = 0\,.$$

In fact

$$\| T_1 - T_2 \| = \max_{0\leq x\leq\pi}\left|\exp[-\sqrt{n}]\sin\frac{nx}{1+\lambda n^2} - 0\right|$$

$$\leq \exp[-\sqrt{n}]\,, \tag{1.109}$$

$$\| T_1 - T_2 \| = \max_{\substack{0\leq x\leq\pi\\ t_0\leq t\leq 0}}\left|\exp\left[-\sqrt{n} - \frac{n^2 t}{1+\lambda n^2}\right]\sin\frac{nx}{1+\lambda n^e}\right|$$

$$\leq \left[-\sqrt{n} - \frac{n^2 t_0}{1+\lambda n^2}\right]\,. \tag{1.110}$$

Hence it follows that at $\lambda \neq 0$, $n \to \infty$, right-hand sides (1.109) and (1.110) vanish and instability disappears. Now calculate $v(x)$ for (1.108)

$$v(x) = \sqrt{\frac{k}{\gamma}}\,n\tan\frac{nx}{1+\lambda n^2}\,. \tag{1.111}$$

If for solution (1.96) obtained from (1.93) there are points where $v = 0$ or $v = \infty$, then for solution (1.108) as seen from (1.111), this peculiarity can be eliminated by a correct choice of λ. It means that heating through the substance from a boundary into the region $0 < x < \pi$ will proceed with a finite nonzero rate in the agreement with special relativity theory.

References

1.1. W. Nernst: *Die Theorethischen Grundlagen des Wärmestatzes*, (Knapp Halle 1917)

1.2. J. Scales, R. Snider: Nature **401**, 739 (1999)

1.3. E. Zauderer: *Partial Diferential Equation of Applied Mathematics* (J. Willey and Sons N.Y. USA 1989)

1.4. J.L. Davis: *Wave Propagation*, (Princeton University Press Princeton USA 2000)

1.5. C.K. Sun et al.: Phys. Rev. **B50**, 15337 (1994)

1.6. N. Del Fatti et al.: Phys. Rev. **B61**, 16956 (2000)

1.7. C. Voisin et al.: Phys. Rev. Lett. **85**, 2200 (2000)

1.8. N. Del Fatti, F. Vallée: C. R. Acad. Sci. Paris **3**, 365 (2002)

1.9. C. Guillon et al.: New Journal of Physics **5**, 131 (2003)

1.10. M. Kozlowski, J. Marciak-Kozlowska: Lasers Eng. **7**, 81 (1998)

1.11. M. Kozlowski, J. Marciak-Kozlowska: Lasers Eng. **9**, 39 (1999)

1.12. M. Kozlowski, J. Marciak-Kozlowska: Lasers Eng. **10**, 293 (2000)

1.13. J.A. Scales, R. Snieder: Nature vol 401, 739 (1999)

1.14. J.P. Wolfe: *Imaging Phonons Acoustic Wave Propagation in Solids*, (Cambridge Univ. Press 1998)

1.15. R. Bellman, R. Kalaba, G. Wing: J. Math. Mech. **9**, 933 (1960)

1.16. O. Narayan, S. Ramaswamy: Phys. Rev. Lett. **89**, 200601-1 (2002)

1.17. J.S. Hadamard: *Le Probléme de Cauchy les Equations aux Dérivées Particelles Lineaires Hyperboliques* (Hermann Paris 1932)

Causal Thermal Phenomena, Classical Description

2.1 Fundamentals of the Rapid Thermal Processes

2.1.1 The Memory Function for Thermal Processes

In paper [2.1] published in 1956, M. Kac considered a particle moving on line at speed c, taking discrete steps of equal size, and undergoing collisions (reversals of direction) at random times, according to a Poisson process of intensity a. He showed that the expected position of the particle satisfies either of two difference equations, according to its initial direction. With correct scaling followed by a passage to the limit, the difference equations become a pair of first-order partial differential equations (PDE). Differentiating those and adding them yields the hyperbolic diffusion equation (telegraph equation)

$$\frac{\partial^2 u}{\partial t^2} + a\frac{\partial u}{\partial t} = c\frac{\partial}{\partial x}\left(c\frac{\partial u}{\partial x}\right). \tag{2.1}$$

This is an equation of hyperbolic type. If the lower term (in time) is dropped, it's just the one-dimensional wave equation.

In paper [2.2], R. Hersh proposed the operator generalization of (2.1):

$$\frac{\partial}{\partial t}\left(\frac{\partial u}{\partial t}\right) + a\frac{\partial u}{\partial t} = A^2 u. \tag{2.2}$$

In equation (2.2), A is the generator of a group of linear operators acting on a linear space B. Instead of transition moving randomly to the left and right at speed c, the time evolution according to generators A and $^*A$ is substituted.

The study and applications of the classical hyperbolic diffusion equation (2.1) covers the thermal processes [2.3, 2.4], stock prices [2.5], astrophysics [2.6], and heavy ion physics [2.7].

In this section, we will study the ultrashort thermal processes in the framework of the hyperbolic diffusion equation.

When an ultrafast thermal pulse (e.g. femtosecond pulse) interacts with a metal surface, the excited electrons become the main carriers of the thermal energy. For a femtosecond thermal pulse, the duration of the pulse is of the same order as the electron relaxation time. In this case, the hyperbolicity of the thermal energy transfer plays an important role.

Radiation deposition of energy in materials is a fundamental phenomenon to laser processing. It converts radiation energy into a material's internal energy, which initiates many thermal phenomena, such as heat pulse propagation, melting, and evaporation. The operation of many laser techniques requires an accurate understanding and control of the energy deposition and transport processes. Recently, radiation deposition and the subsequent energy transport in metals have been investigated with picosecond and femtosecond resolutions [2.8]–[2.14]. Results show that during high-power and short-pulse laser heating, free electrons can be heated to an effective temperature much higher than the lattice temperature, which in turn leads to both a much faster energy propagation process and a much smaller lattice-temperature rise than those predicted from the conventional radiation heating model. Corkum et al. [2.15] found that this electron-lattice nonequilibrium heating mechanism can significantly increase the resistance of molybdenum and copper mirrors to thermal damage during high-power laser irradiation when the laser pulse duration is shorter than one nanosecond. Clemens et al. [2.16] studied thermal transport in multilayered metals during picosecond laser heating. The measured temperature response in the first 20 ps was found to be different from

Table 2.1. General Features of Heat Carriers

	Free Electron	Phonon
Generation	Ionization or excitation	Lattice vibration
Propagation media	Vacuum or media	Media only
Statistics	Fermion	Boson
Dispersion	$E = \hbar^2 q^2/(2m)$	$E = E(q)$
Velocity (m/s)	$\sim 10^6$	$\sim 10^3$

predictions of the conventional Fourier model. Due to the relatively low temporal resolution of the experiment (~ 4 ps), however, it is difficult to determine
whether this difference is the result of nonequilibrium laser heating or is due
to other heat conduction mechanisms, such as non-Fourier heat conduction,
or reflection and refraction of thermal waves at interfaces.

Heat is conducted in solids through electrons and phonons. In metals, electrons dominate the heat conduction, while in insulators and semiconductors,
phonons are the major heat carriers. Table 2.1 lists important features of the
electrons and phonons.

The traditional thermal science, or macroscale heat transfer, employs phe-
nomenological laws, such as Fourier's law, without considering the detailed
motion of the heat carriers. Decreasing dimensions, however, have brought an
increasing need for understanding the heat transfer processes from the microscopic point of view of the heat carriers. The response of the electron and
phonon gases to the external perturbation initiated by laser irradiation can
be described with the help of a memory function of the system. To that aim,
let us consider the generalized Fourier law [2.17]–[2.20]:

$$q(t) = - \int_{-\infty}^{t} K(t - t') \nabla T(t') \, dt', \qquad (2.3)$$

where $q(t)$ is the density of a thermal energy flux, $T(t')$ is the temperature of
electrons, and $K(t - t')$ is a memory function for thermal processes. The density of thermal energy flux satisfies the following equation of heat conduction:

$$\frac{\partial}{\partial t} T(t) = \frac{1}{\varrho c_v} \nabla^2 \int_{-\infty}^{t} K(t - t') T(t') \, dt', \qquad (2.4)$$

where ϱ is the density of charge carriers, and c_v is the specific heat of electrons
in a constant volume. We introduce the following equation for the memory
function describing the Fermi gas of charge carriers:

$$K(t - t') = K_1 \lim_{t_0 \to 0} \delta(t - t' - t_0). \qquad (2.5)$$

In this case, the electron has a very "short" memory due to thermal disturbances of the state of equilibrium. Combining (2.5) and (2.4) we obtain

$$\frac{\partial}{\partial t} T = \frac{1}{\varrho c_v} K_1 \nabla^2 T. \qquad (2.6)$$

Equation (2.6) has the form of the parabolic equation for heat conduction (PHC). Using this analogy, (2.6) may be transformed as follows:

$$\frac{\partial}{\partial t}T = D_T \nabla^2 T, \tag{2.7}$$

where the heat diffusion coefficient D_T is defined as follows:

$$D_T = \frac{K_1}{\varrho c_v}. \tag{2.8}$$

From (2.8), we obtain the relation between the memory function and the diffusion coefficient

$$K(t - t') = D_T \varrho c_v \lim_{t_0 \to 0} \delta(t - t - t_0'). \tag{2.9}$$

In the case when the electron gas shows a "long" memory due to thermal disturbances, one obtains for memory function

$$K(t - t') = K_2. \tag{2.10}$$

When (2.10) is substituted into (2.4), we obtain

$$\frac{\partial}{\partial t}T = \frac{K_2}{\varrho c_v} \nabla^2 \int_{-\infty}^{t} T(t') \, dt. \tag{2.11}$$

Differentiating both sides of (2.11) with respect to t, we obtain

$$\frac{\partial^2 T}{\partial t^2} = \frac{K_2}{\varrho c_v} \nabla^2 T. \tag{2.12}$$

Equation (2.12) is the hyperbolic wave equation describing thermal wave propagation in a charge carrier gas in a metal film. Using a well-known form of the wave equation,

$$\frac{1}{v^2}\frac{\partial^2 T}{\partial t^2} = \nabla^2 T, \tag{2.13}$$

and comparing (2.12) and (2.13), we obtain the following form for the memory function:

$$K(t - t') = \varrho c_v v^2 \tag{2.14}$$

$$v = \text{finite}, \ v < \infty.$$

As the third case, "intermediate memory" will be considered:

$$K(t - t') = \frac{K_3}{\tau} \exp\left[-\frac{(t - t')}{\tau}\right], \tag{2.15}$$

where τ is the relaxation time of thermal processes. Combining (2.15) and (2.4), we obtain

$$c_v \frac{\partial^2 T}{\partial t^2} + \frac{c_v}{\tau}\frac{\partial T}{\partial t} = \frac{K_3}{\varrho \tau} \nabla^2 T \tag{2.16}$$

and

$$K_3 = D_\tau c_v \varrho. \tag{2.17}$$

Thus, finally,

$$\frac{\partial^2 T}{\partial t^2} + \frac{1}{\tau} \frac{\partial T}{\partial t} = \frac{D_T}{\tau} \nabla^2 T. \tag{2.18}$$

Equation (2.18) is the hyperbolic equation for heat conduction (HHC), in which the electron gas is treated as a Fermion gas. The diffusion coefficient D_T can be written in the form [2.21]

$$D_T = \frac{1}{3} v_F^2 \tau, \tag{2.19}$$

where v_F is the Fermi velocity for the electron gas in a semiconductor. Applying (2.19), we can transform the hyperbolic equation for heat conduction, (2.18), as follows:

$$\frac{\partial^2 T}{\partial t^2} + \frac{1}{\tau} \frac{\partial T}{\partial t} = \frac{1}{3} v_F^2 \nabla^2 T. \tag{2.20}$$

Let us denote the velocity of disturbance propagation in the electron gas as s:

$$s = \sqrt{\frac{1}{3} v_F}. \tag{2.21}$$

Using the definition of s, (2.20) may be written in the form

$$\frac{1}{s^2} \frac{\partial^2 T}{\partial t^2} + \frac{1}{\tau s^2} \frac{\partial T}{\partial t} = \nabla^2 T. \tag{2.22}$$

For the electron gas, treated as the Fermi gas, the velocity of sound propagation is described by equation (2.22)

$$v_S = \left(\frac{P_F^2}{3mm^*} \left(1 + F_0^S\right) \right)^{1/2}, \qquad P_F = m v_F, \tag{2.23}$$

where m is the mass of a free (noninteracting) electron and m^* is the effective electron mass. Constant F_0^S represents the magnitude of carrier-carrier interaction in the Fermi gas. In the case of a very weak interaction, $m^* \to m$ and $F_0^S \to 0$, so according to (2.23),

$$v_S = \frac{m v_F}{\sqrt{3m}} = \sqrt{\frac{1}{3} v_F}. \tag{2.24}$$

To sum up, we can make a statement that for the case of weak electron-electron interaction, sound velocity $v_s = \sqrt{1/3} v_F$, and this velocity is equal to the velocity of thermal disturbance propagation s. From this, we conclude that the hyperbolic equation for heat conduction (2.22) is identical as the equation for second sound propagation in the electron gas:

$$\frac{1}{v_S^2} \frac{\partial^2 T}{\partial t^2} + \frac{1}{\tau v_S^2} \frac{\partial T}{\partial t} = \nabla^2 T. \tag{2.25}$$

Using the definition expressed by (2.19) for the heat diffusion coefficient, (2.25) may be written in the form

$$\frac{1}{v_S^2} \frac{\partial^2 T}{\partial t^2} + \frac{1}{D_T} \frac{\partial T}{\partial t} = \nabla^2 T. \tag{2.26}$$

The mathematical analysis of (2.25) leads to the following conclusions:

1. In the case when $v_S^2 \to \infty$, τv_S^2 is finite, (2.26) transforms into the parabolic equation for heat diffusion:

$$\frac{1}{D_T} \frac{\partial T}{\partial t} = \nabla^2 T. \tag{2.27}$$

2. In the case when $\tau \to \infty$, v_S is finite, (2.26) transforms into the wave equation:

$$\frac{1}{v_S^2} \frac{\partial^2 T}{\partial t^2} = \nabla^2 T. \tag{2.28}$$

Equation (2.28) describes propagation of the thermal wave in the electron gas. From the point of view of theoretical physics, condition $v_S \to \infty$ violates the special theory of relativity. From this theory, we know that there is a limited velocity of interaction propagation and this velocity $v_{\lim} = c$, where c is the velocity of light in a vacuum. Multiplying both sides of (2.26) by c^2, we obtain

$$\frac{c^2}{v_S^2} \frac{\partial^2 T}{\partial t^2} + \frac{c^2}{D_T} \frac{\partial T}{\partial t} = c^2 \nabla^2 T, \tag{2.29}$$

Denoting $\beta = v_S/c$, (2.29) may be written in the form

$$\frac{1}{\beta^2} \frac{\partial^2 T}{\partial t^2} + \frac{1}{\tilde{D}_T} \frac{\partial T}{\partial t} = c^2 \nabla^2 T, \tag{2.30}$$

where $\tilde{D}_T = \tau \beta^2$, $\beta < 1$. On the basis of the above considerations, we conclude that the heat conduction equation, which satisfies the special theory of relativity, acquires the form of the partial hyperbolic equation (2.30). The rejection of the first component in (2.30) violates the special theory of relativity.

2.1.2 The Relaxation Dynamics of the Ultrafast Thermal Pulses

Heat transport during fast laser heating of solids has become a very active research area due to the significant applications of short pulse lasers in the fabrication of sophisticated microstructures, synthesis of advanced materials, and measurements of thin film properties. Laser heating of metals involves the deposition of radiation energy on electrons, the energy exchange between electrons and the lattice, and the propagation of energy through the media.

The theoretical predictions showed that under ultrafast excitation conditions, the electrons in a metal can exist out of equilibrium with the lattice for times of the order the electron energy relaxation time [2.9, 2.12]. Model calculations suggest that it should be possible to heat the electron gas to temperature T_e of up to several thousand degrees for a few picoseconds while keeping the lattice temperature T_l relatively cold. Observing the subsequent equilibration of the electronic system with the lattice allows one to directly study electron-phonon coupling under various conditions.

Several groups have undertaken investigations relating dynamics' changes in the optical constants (reflectivity, transmissivity) to relative changes in electronic temperature. But only recently, the direct measurement of electron temperature has been reported.

The temperature of hot electron gas in a thin gold film ($l = 300$ Å) was measured, and a reproducible and systematic deviation from a simple Fermi-Dirac (FD) distribution for short time $\Delta t \sim 0.4\,\mathrm{ps}$ was obtained. As stated in [2.12], this deviation arises due to the finite time required for the nascent electrons to equilibrate to a FD distribution. The nascent electrons are the electrons created by the direct absorption of the photons prior to any scattering.

In papers [2.17, 2.20], the relaxation dynamics of the electron temperature with the hyperbolic heat transport equation (HHT), (2.26), was investigated. Conventional laser heating processes that involve a relatively low-energy flux and long laser pulse have been successfully modeled in metal processing and in measuring thermal diffusivity of thin films [2.23]. However, applicability of these models to short-pulse laser heating is questionable [2.9, 2.12, 2.17, 2.18, 2.19, 2.20]. As it is well-known, the Anisimov model [2.23] does not properly take into account the finite time for the nascent electrons to relax to the FD distribution. In the Anisimov model, the Fourier law for heat diffusion in the electron gas is assumed. However, the diffusion equation is valid only when relaxation time is zero, $\tau = 0$, and velocity of the thermalization is infinite, $v \to \infty$.

The effects of ultrafast heat transport can be observed in the results of front-pump back probe measurements [2.9, 2.12]. The results of these type of experiments can be summarized as follows. Firstly, the measured delays are much shorter than would be expected if heat were carried by the diffusion of electrons in equilibrium with the lattice (tens of picoseconds). This suggests that heat is transported via the electron gas alone, and that the electrons are out of equilibrium with the lattice on this time scale. Secondly, because the delay increases approximately linearly with the sample thickness, the heat transport velocity can be extracted, $v_h \simeq 10^8 \, \text{cm/s} = 1 \, \mu\text{m/ps}$. This is of the same order of magnitude as the Fermi velocity of electrons in gold, $1.4 \, \mu\text{m/ps}$.

2.1.3 The Thermal Inertia of Materials Heated with Ultrafast Laser Pulses

According to the constitutive relation in the thermal wave model, heat flux q obeys the relation [2.17]–[2.20]

$$q(r, t + \tau) = -k\nabla T(r, t),$$ (2.31)

where τ is the relaxation time (a phase lag) and k is the thermal conductivity. The temperature gradient established in the material at time t results in a heat flux that occurred at a later time $t + \tau$ due to the insufficient time of response. For combining with the energy equation, however, all the physical quantities involved must correspond to the same instant of time. The Taylor's series expansion is thus applied to the heat flux q in (2.31) to give

$$q(r, t) + \frac{\partial q(r, t)}{\partial t} \tau + \frac{\partial^2 q(r, t)}{\partial t^2} \frac{\tau^2}{2} = -k\nabla T(r, t).$$ (2.32)

In the linearized thermal wave theory, the phase lag is assumed to be small and the higher - order terms in (2.32) are neglected. By retaining only the first-order term in τ, (2.32) becomes

$$q(r, t) + \tau \frac{\partial q(r, t)}{\partial t} = -k\nabla T(r, t).$$ (2.33)

After combining (2.33) with the energy conservation equation

$$-\nabla \cdot q = \varrho C_v \frac{\partial T}{\partial t}$$ (2.34)

one obtains the HHT, (2.26).

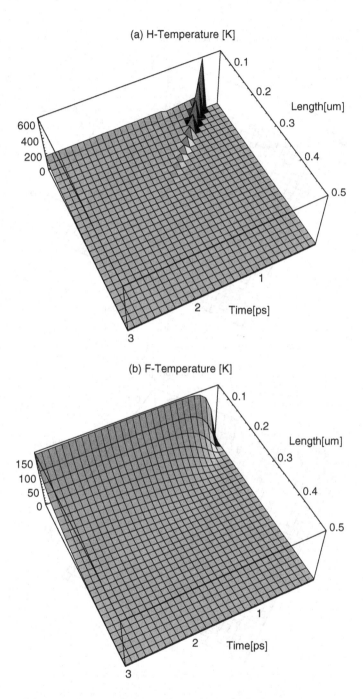

Fig. 2.1. (a) The solution of HHT for $v_S = 0.15\,\mu\text{m/ps}$, $\tau = 0.12\,\text{ps}$, Δt-pulse duration $= 0.06\,\text{ps}$. (b) The solution of PHT for the same values of v_S, τ, and $\Delta t = 0.06\,\text{ps}$.

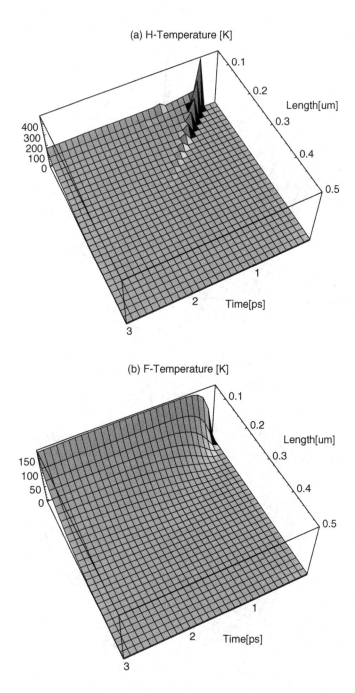

Fig. 2.2. (a) The same as in Fig. 2.1(a) but with Δt-pulse duration = 0.1 ps. (b) The same as in Fig. 2.1(b), but with $\Delta t = 0.1$ ps.

Equation (2.33) can be compared with the equation of the motion for particle with mass m in a resistive medium,

$$\gamma \boldsymbol{v} + m \frac{d\boldsymbol{v}}{dt} = \boldsymbol{P}(\boldsymbol{r}, t), \qquad (2.35)$$

where γ is a resistive coefficient, $\boldsymbol{v}$ denotes the velocity, and $\boldsymbol{P}(\boldsymbol{r}, t)$ is the external force. Comparing (2.33) and (2.35), we conclude the correspondence

$$- \nabla T(\boldsymbol{r}, t) \longrightarrow \boldsymbol{P}(\boldsymbol{r}, t),$$

$$\boldsymbol{q}(\boldsymbol{r}, t) \longrightarrow \boldsymbol{v},$$

$$k^{-1} \longrightarrow \gamma,$$

$$\frac{\tau}{k} \longrightarrow m. \qquad (2.36)$$

For the steady-state case, (2.35) reduces to

$$\gamma \boldsymbol{v} = \boldsymbol{P}(\boldsymbol{r}, t) \qquad (2.37)$$

and (2.32) reduces to the Fourier law.

From the relations, given by (2.36), we conclude that the greater relaxation time corresponds to the greater mass $\equiv$ greater inertia. It seems quite reasonable to treat the relaxation time as the measure of the degree of the thermal inertia.

In papers [2.17]–[2.20], it was shown that for the thermal processes with characteristic time $\Delta t \geq \tau$, the heat transfer is well described by Fourier law. In another way, for $\Delta t \geq \tau$ the thermal processes can be called inertia-free processes. On the other hand, for thermal processes with $\Delta t < \tau$, the thermal inertia plays an important role.

In Figs. 2.1 and 2.2, the 3D solutions of HHT and PHT (Parabolic Heat Equation) equations are presented. The solutions are obtained for $v_S = 0.15\,\mu\mathrm{m/ps}$ and $\tau = 0.12\,\mathrm{ps}$ [2.20] and for $\Delta t = 0.06\,\mathrm{ps}$ and $0.1\,\mathrm{ps}$. As can be easily seen, the solutions of HHT equations (Figs. 2.1(a) and 2.2(a)), show the retardation of the response of the system to the external thermal perturbation. The temperature surface shows the effect of the thermal inertia. Moreover, the shorter the Δt, the more localized is the temperature surface. For the solution of PHT (Figs. 2.1(b) and 2.2(b)) the instant heating of the system is observed without any signature of the inertia of the system, and the temperature in system is smeared out.

2.1.4 Causal Transport of Hot Electrons

The effects of ultrafast heat transport can be observed in the results of front-pump back probe measurements [2.12]. The results of these type of experiments can be summarized as follows: Firstly, the measured delays are much shorter than would be expected if the heat were carried by the diffusion of electrons in equilibrium with the lattice (tens of picoseconds). This suggests that the heat is transported via the electron gas alone and that the electrons are out of equilibrium with the lattice on this time scale. Secondly, because the delay increases approximately linearly with the sample thickness, the heat transport velocity can be extracted $v_h \simeq 10^8 \, \text{cm/s} = 1 \, \mu\text{m/ps}$. This is of the same order of magnitude as the Fermi velocity of electrons in Au, $1.4 \, \mu\text{m/ps}$.

Because the heat moves at a velocity comparable to v_F (Fermi velocity) of the electron gas, it is natural to question exactly how the transport takes place. Because those electrons that lie close to the Fermi surface are the principal contributors to transport, the heat-carrying electrons move at v_F. In the limit of lengths longer than the momentum relaxation length, λ, the random walk behavior is averaged and the electron motion is subjected to a diffusion equation. Conversely, on a length scale shorter than λ, the electrons move ballistically with velocity close to v_F.

The importance of the ballistic motion may be appreciated by considering the different hot-electron scattering lengths reported in the literature. The electron-electron scattering length in Au, λ_{ee}, has been calculated in [2.24]. They find that $\lambda_{ee} \sim (E - E_F)^2$ for electrons close to the Fermi level. For 2-eV electrons, $\lambda_{ee} \approx 35 \, \text{nm}$ increasing to 80 nm for 1 eV. The electron-phonon scattering length λ_{ep} is usually inferred from conductivity data. Using Drude relaxation times, λ_{ep} can be computed, $\lambda_{ep} \approx 42 \, \text{nm}$ at 273 K. This is shorter than λ_{ee} but of the same order of magnitude. Thus, we would expect that both electron-electron and electron-phonon scattering are important on this length scale. However, because conductivity experiments are steady-state measurements, the contribution of phonon scattering in a femtosecond regime experiment such as pump-probe ultrafast lasers is uncertain.

In the usual electron-phonon coupling model [2.23], one describes the metal as two coupled subsystems, one for electrons and one for phonons. Each subsystem is in local equilibrium so the electrons are characterized by a FD distribution at temperature T_e, and the phonon distribution is characterized by a Bose-Einstein distribution at the lattice temperature T_l. The coupling between the two systems occurs via the electron-phonon interaction. The time evolution of the energies in the two subsystems is given by the coupled parabolic differential equations (Fourier law). For ultrafast lasers, the duration of pump pulse is of the order relaxation time in metals [2.12]. In that case, the parabolic heat conduction equation is not valid, and hyperbolic heat transport equation must be used (2.26)

$$\frac{1}{v_S^2} \frac{\partial^2 T}{\partial t^2} + \frac{1}{D_T} \frac{\partial T}{\partial t} = \nabla^2 T, \qquad D_T = \tau v_S^2. \qquad (2.38)$$

In equation (2.38), v_S is the thermal wave speed, τ is the relaxation time, and D_T denotes the thermal diffusivity. In the following, (2.38) will be used to describe the heat transfer in the thin gold films.

To that aim, we define: T_e is the electron gas temperature and T_l is the lattice temperature. The governing equations for nonstationary heat transfer are

$$\frac{\partial T_e}{\partial t} = D_T \nabla^2 T - \frac{D_T}{v_S^2} \frac{\partial^2 T_e}{\partial t^2} - G(T_e - T_l), \qquad \frac{\partial T_l}{\partial t} = G(T_e - T_l), \quad (2.39)$$

where D_T is the thermal diffusivity, T_e is the electron temperature, T_e is the lattice temperature, and G is the electron-phonon coupling constant. In the following, we will assume that on sub-picosecond scale, the coupling between electron and lattice is weak, and (2.39) can be replaced by the following equations (2.26):

$$\frac{\partial T_e}{\partial t} = D_T \nabla^2 T - \frac{D_T}{v_S^2} \frac{\partial^2 T_e}{\partial t^2}, \qquad T_l = \text{constant}. \qquad (2.40)$$

Equation (2.40) describes nearly ballistic heat transport in a thin gold film irradiated by an ultrafast ($\Delta t < 1\,\mathrm{ps}$) laser beam. The solution of (2.40) for $1\,\mathrm{D}$ is given by [2.17]–[2.20]:

$$T(x,t) = \frac{1}{v_s} \int \mathrm{d}x' \, T(x',0) \left[e^{-t/2\tau} \frac{1}{t_0} \Theta(t - t_0) \right.$$

$$+ e^{-t/2\tau} \frac{1}{2\tau} \left\{ I_0 \left(\frac{(t^2 - t_0^2)^{1/2}}{2\tau} \right) \right. \qquad (2.41)$$

$$\left. \left. + \frac{t}{(t^2 - t_0^2)^{1/2}} I_1 \left(\frac{(t^2 - t_0^2)^{1/2}}{2\tau} \right) \right\} \Theta(t - t_0) \right],$$

where v_s is the velocity of second sound, $t_0 = (x - x')/v_s$, I_0 and I_1 are modified Bessel functions, and $\Theta(t-t_0)$ denotes the Heaviside function. We are concerned with the solution to (2.41) for a nearly delta function temperature pulse generated by laser irradiation of the metal surface. The pulse transferred to the surface has the shape:

$$\Delta T_0 = \frac{\beta \varrho_E}{C_V v_s \Delta t} \qquad \text{for} \qquad 0 \le x < v_s \Delta t,$$

$$\Delta T_0 = 0 \qquad \text{for} \qquad x \ge v_s \Delta t. \tag{2.42}$$

In equation (2.42), ϱ_E denotes the heating pulse fluence, β is the efficiency of the absorption of energy in the solid, $C_V(T_e)$ is electronic heat capacity, and Δt is duration of the pulse. With $t = 0$, temperature profile described by (2.42) yields:

$$T(l, t) = \frac{1}{2} \Delta T_0 e^{-t/2\tau} \Theta(t - t_0)\Theta(t_0 + \Delta t - t) \tag{2.43}$$

$$+ \frac{\Delta t}{4\tau} \Delta T_0 e^{-\tau/2\tau} \left\{ I_0(z) + \frac{t}{2\tau} \frac{1}{z} I_1(z) \right\} \Theta(t - t_0),$$

where $z = (t^2 - t_0^2)^{1/2}/2\tau$ and $t = l/v_s$. The solution to (2.40), when there are reflecting boundaries, is the superposition of the temperature at l from the original temperature and from image heat source at $\pm 2nl$. This solution is

$$T(l, t) = \sum_{i=0}^{\infty} \Delta T_0 e^{-t/2\tau} \Theta(t - t_i)\Theta(t_i + \Delta t - t) \tag{2.44}$$

$$+ \Delta T_0 \frac{\Delta t}{2\tau} e^{-t/2\tau} \left\{ I_0(z_i) + \frac{t}{2\tau} \frac{1}{z_i} I_1(z_i) \right\} \Theta(t - t_i),$$

where $t_i = t_0, 3t_0, 5t_0$, $t_0 = l/v_0$. For gold, $C_V(T_e) = C_e(T_e) = \gamma T_e$, $\gamma = 71.5 \, \text{J}/(\text{m}^3\text{K}^2)$ and (2.42) yields:

$$\Delta T_0 = \frac{1.4 \, 10^5 \, \varrho_E \beta}{v_s \Delta t T_e} \qquad \text{for} \qquad 0 \le x \le v_s \Delta t,$$

$$\Delta T_0 = 0 \qquad \text{for} \qquad x \ge v_s \Delta t, \tag{2.45}$$

where ϱ_E is measured in $\text{m J}/\text{cm}^2$, v_s in $\mu\text{m}/\text{ps}$, and Δt in ps. For $T_e = 300 \, \text{K}$:

$$\Delta T_0 = \frac{4.67 \, 10^2 \beta \varrho_E}{v_s \Delta t} \qquad \text{for} \qquad 0 \le x \le v_s \Delta t,$$

$$\Delta T_0 = 0 \qquad \text{for} \qquad x \ge v_s \Delta t. \tag{2.46}$$

The model calculations (formulae 2.43–2.46) were applied to the description of the experimental results presented in paper [2.12], and a fairly good agreement of the theoretical calculations and experimental results was obtained [2.20].

2.2 The Thermal Wave as the Solution of HHT

2.2.1 Velocity of Thermal Waves

In the early 1950s, it was shown by Dingle [2.26], Ward and Wilks [2.27], and London [2.28] that a density fluctuation in a phonon gas would propagate as a thermal wave – a second sound wave – provided that "losses" from the wave were negligible. In one of their papers, Ward and Wilks [2.29] indicated they would attempt to look for a second sound wave in sapphire crystals. No results of their experiments were published. Then, for nearly a decade, the subject of "thermal wave" lay dormant. Interest was revived in the 1960s, primarily through the efforts of J.A. Krumhansl, R.A. Guyer, and C.C. Ackerman. In the paper by Ackerman and Guyer [2.30], the thermal wave in dielectric solids was experimentally and theoretically investigated. They found a value for the thermal wave velocity in LiF at a very low temperature $T \sim 1$ K, of $v_s \sim 100 - 300$ m/s. In insulators and semiconductors, phonons are the major heat carriers. In metals, electrons dominate. For long thermal pulses, i.e., when the pulse duration, Δt, is larger than the relaxation time, τ, for thermal processes, $\Delta t \gg \tau$, the heat transfer in metals is well described by Fourier diffusion equation. The advent of modern ultrafast lasers opens up the possibility of investigating a new mechanism of thermal transport – the thermal wave in an electron gas heated by lasers. The effect of an ultrafast heat transport can be observed in the results of front pump back probe measurements [2.9, 2.12]. The results of this type of experiment can be summarized as follows. Firstly, the measured delays are much shorter than it would be expected if the heat were carried by the diffusion of electrons in equilibrium with the lattice (tens of picoseconds). This suggests that the heat is transported via the electron gas alone and that the electrons are out of equilibrium with the lattice within this time scale. Secondly, because the delay increases approximately linearly with the sample thickness, the heat transport velocity can be determined, $v_h \sim 10^8$ cm/s $= 1\mu$m/ps. This is of the same order of magnitude as the Fermi velocity of electrons in Au, 1.4 μm/ps.

In the papers [2.9, 2.19], the heat transport in a thin metal film (Au) was investigated with the help of the hyperbolic heat conduction equation. It was shown that when the memory of the hot electron gas in metals is taken into account, then the HHT is the dominant equation for heat transfer. The hyperbolic heat conduction equation for heat transfer in an electron gas has the form (2.26)

$$\frac{1}{\left(\frac{1}{3}v_F^2\right)}\frac{\partial^2 T}{\partial t^2} + \frac{1}{\tau\left(\frac{1}{3}v_F^2\right)}\frac{\partial T}{\partial t} = \nabla^2 T\,. \tag{2.47}$$

If we consider an infinite electron gas, then the Fermi velocity can be calculated

$$v_F \cong bc\,. \tag{2.48}$$

In equation (2.48), c is the light velocity in vacuum and $b \sim 10^{-2}$. Considering (2.48), (2.47) can be written in a more elegant form:

$$\frac{1}{c^2}\frac{\partial^2 T}{\partial t^2} + \frac{1}{c^2\tau}\frac{\partial T}{\partial t} = \frac{b^2}{3}\nabla^2 T \tag{2.49}$$

In order to derive the Fourier law from (2.49), we are forced to break the special theory of relativity and put in (2.49) $c \to \infty$, $\tau \to 0$. In addition, it can be demonstrated from HHT in a natural way that in electron gas, the heat propagation with velocity $v_h \sim v_F$ is the accordance with the results of the pump probe experiments [2.30, 2.9].

2.2.2 The Thermal Wave as the Solution of HHT

The importance of the existence of thermal waves in engineering applications was investigated in the paper [2.19]. The propagation of the thermal front in metals has attracted a lot of attention and presents a unique feature of thermal wave propagation.

Considering the importance of the thermal wave in future engineering applications and simultaneously the lack of the simple physics presentation of the thermal wave for engineering audience, in the following we present the main results concerning the wave nature of heat transfer.

Hence, we discuss (2.49) in more detail. Firstly, we observe that the second derivative term dominates when:

$$c^2(\Delta t)^2 < c^2\Delta t\tau \tag{2.50}$$

i.e., when $\Delta t < \tau$. This implies that for very short heat pulses, we have a hyperbolic wave equation of the form:

$$\frac{1}{c^2}\frac{\partial^2 T}{\partial t^2} = \frac{b^2}{3}\nabla^2 T \tag{2.51}$$

and the velocity of the thermal wave is given by

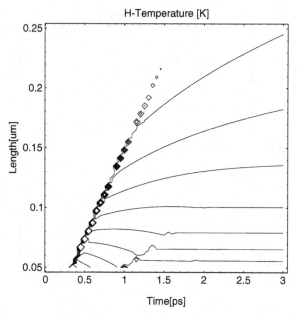

Fig. 2.3. The solution of the HHT (2.49) "H-Temperature" for thermal wave velocity $v_{th} = 0.15$ μm/ps, relaxation time $\tau = 0.1$ ps, and pulse duration $\Delta t = 0.02$ ps.

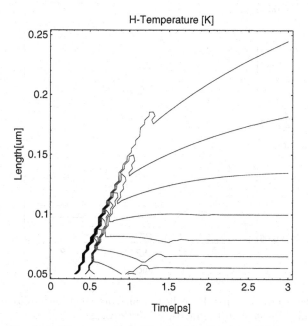

Fig. 2.4. The same as in Fig. 2.3 but for $\Delta t = 0.06$ ps.

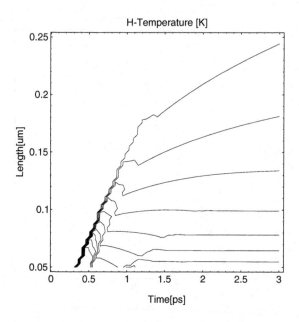

Fig. 2.5. The same as in Fig. 2.3 but for $\Delta t = 0.1$ ps.

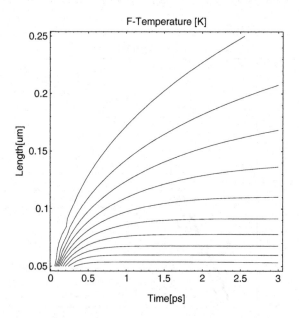

Fig. 2.6. The solution of parabolic heat transport equation (2.55) Fourier law. "F-Temperature" for $v_{th} = 0.15$ μm/ps, $\tau = 0.12$ ps, $\Delta t = 0.02$ ps.

$$v_{th} \sim \frac{1}{\sqrt{3}} \frac{c}{b}, \qquad b \sim 10^{-2}. \tag{2.52}$$

The velocity v_{th} (2.52) is the velocity of the thermal wave in an infinite Fermi gas of electrons, which is free of all impurities. The thermal wave, which is described by the solution of (2.51), does not interact with the crystal lattice. It is the maximum value of the thermal wave obtainable in an infinite free electron gas.

If we consider the opposite case to that in (2.50)

$$c^2(\Delta t)^2 > c^2 \tau \Delta t \tag{2.53}$$

i.e., when

$$\Delta t > \tau \tag{2.54}$$

then, one obtains from (2.49):

$$\frac{1}{c^2 \tau} \frac{\partial T}{\partial t} = \frac{b^2}{3} \nabla^2 T . \tag{2.55}$$

Equation (2.55) is the parabolic heat conduction equation–Fourier equation.

The solutions of (2.49) for the following input parameters: $\tau = 0.12$ ps, $v_{th} = 0.15$ μm/ps, $\Delta t = 0.02$ ps, 0.06 ps, 0.1 ps are presented in Figs. 2.3–2.5.

The value of the thermal wave velocity v_h is taken from paper [2.20]. Isotherms are presented as a function of the thin film thickness (length) l [μm] and the delay times. The mechanism of heat transfer on a nanometer scale can be divided according to Fig. 2.3 ($\Delta t = 0.02$ ps) into three stages: a heat wave for $t \sim Lv_{th}^{-1}$, mixed heat transport for $Lv_{th}^{-1} < t < 3L_{th}^{-1}$, and diffusion for $t > 3Lv_{th}^{-1}$. The thermal wave moves in a manner described by the hyperbolic differential partial equation, $x = v_{th} t$. For $t < xv_{th}^{-1}$, the system is undisturbed by an external heat source (laser beam). For longer heat pulses, Figs. 2.4–2.5, the evidence of the thermal wave is gradually reduced – but the retardation of the thermal pulse is still evident. In Fig. 2.6, the solution of the parabolic heat conduction (2.55) is presented. In this case, contrary to the solution of the HHT, heating of the film starts at $t = 0$.

2.2.3 Slowing and Dephasing of the Thermal Waves

If heat is released in a body of gas, liquid, or solid, a thermal flux transported by heat conduction appears. The pressure gradients associated with the thermal gradients set a gas or liquid in motion, so that additional energy transport occurs through convection. In particular, at sufficiently large energy releases, shock waves are formed in a gas or liquid, which transport thermal energy at velocities larger that the speed of sound. Below the critical energy release, nearly pure thermal wave may propagate owing to heat conduction in a gas or liquid with other transport mechanisms being negligible [2.31]. Solid metals provide an ideal test medium for the study of thermal waves, because they are practically incompressible at temperature below their melting point and the thermal wave pressures are small compared with the classic pressure (produced by repulsion of the atoms in the lattice) up to large energy releases. In accordance with this picture, the speed of sound in a metal is independent of temperature and given by $c_s = (E/\varrho)^{1/2}$ where E is the elasticity modulus and ϱ is the density.

Using the path-integral method developed in paper [2.32], the solution of the HHT can be obtained. It occurs that the velocity of the thermal wave in medium is lower than the velocity of the initial thermal wave. The slowing of the thermal wave is caused by the scattering of heat carriers in medium. The scatterings also change the phase of the initial thermal wave.

In one-dimensional flow of heat in metals, the hyperbolic heat transport equation is given by (2.20)

$$\tau \frac{\partial^2 T}{\partial t^2} + \frac{\partial T}{\partial t} = D_T \frac{\partial^2 T}{\partial x^2}, \qquad D_T = \frac{1}{3} v_F^2 \tau, \tag{2.56}$$

where τ denotes the relaxation time, D_T is the diffusion coefficient, and T is the temperature. Introducing the nondimensional spatial coordinate $z = x/\lambdabar$, where $\lambdabar = \lambda/2\pi$ denotes the reduced mean free path, (2.56) can be written in the form:

$$\frac{1}{v'^2} \frac{\partial^2 T}{\partial t^2} + \frac{2a}{v'^2} \frac{\partial T}{\partial t} = \frac{\partial^2 T}{\partial z^2} \tag{2.57}$$

where

$$v' = \frac{v}{\lambdabar} \qquad a = \frac{1}{2\tau} \tag{2.58}$$

In equation (2.58), v denotes the velocity of heat propagation [2.10], $v = (D/\tau)^{1/2}$.

In the paper by C. De Witt-Morette and See Kit Fong [2.32], the path-integral solution of (2.57) was obtained. It was shown that for the initial condition of the form:

$$T(z,0) = \Phi(z) \qquad \text{an ``arbitrary'' function,}$$

$$\left. \frac{\partial T(z,t)}{\partial t} \right|_{t=0} = 0, \tag{2.59}$$

the general solution of the (2.56) has the form

$$T(z,t) = \frac{1}{2} \left[\Phi(z,t) + \Phi(z,-t) \right] e^{-at}$$

$$+ \frac{a}{2} e^{-at} \int_0^t d\eta \left[\Phi(z,\eta) + \Phi(z,-\eta) \right] \tag{2.60}$$

$$+ \left[I_0(a(t^2 - \eta^2)^{1/2}) + \frac{t}{(t^2 - \eta^2)^{1/2}} I_1(a(t^2 - \eta^2)^{1/2}) \right].$$

In equation (2.60), $I_0(x)$ and $I_1(x)$ denote the modified Bessel function of zero and first order, respectively.

Let us consider the propagation of the initial thermal wave with velocity v', i.e.,

$$\Phi(z - v't) = \sin(z - v't). \tag{2.61}$$

In that case, the integral in (2.60) can be computed analytically, $\Phi(z, t) + \Phi(z, -t) = 2 \sin z \cos (v' t)$, and the integrals on the right-hand side of (2.60) can be done explicitly [2.32]; we obtain:

$$F(z,t) = e^{-at} \left[\frac{a}{w_1} \sin (w_1 t) + \cos (w_1 t) \right] \sin z, \qquad v' \geq a \tag{2.62}$$

and

$$F(z,t) = e^{-at} \left[\frac{a}{w_2} \sin h(w_2 t) + \cos h(w_2 t) \right] \sin z, \qquad v' < a \tag{2.63}$$

where $w_1 = (v'^2 - a^2)^{1/2}$ and $w_2 = (a^2 - v'^2)^{1/2}$.

In order to clarify the physical meaning of the solutions given by formulas (2.62) and (2.63), we observe that $v' = v/\lambda$, and w_1 and w_2 can be written as:

$$v_1 = \lambda w_1 = v \left(1 - \left(\frac{1}{2\tau\omega} \right)^2 \right)^{1/2}, \qquad 2\tau\omega > 1$$

$$v_2 = \lambda w_2 = v \left(\left(\frac{1}{2\tau\omega} \right)^2 - 1 \right)^{1/2}, \qquad 2\tau\omega < 1 \tag{2.64}$$

where ω denotes the pulsation of the initial thermal wave. From formula (2.64), it can be concluded that we can define the new effective thermal wave velocities v_1 and v_2.

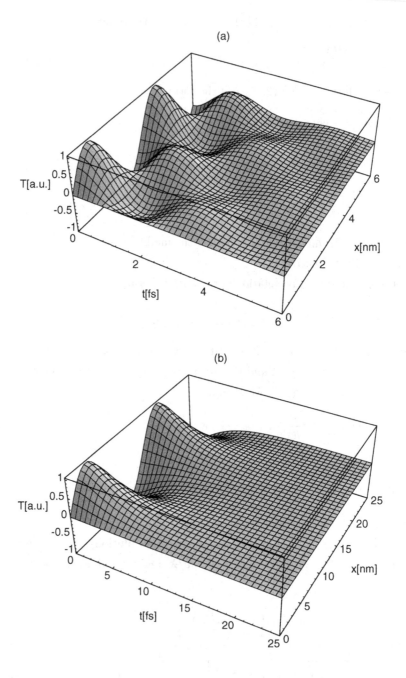

Fig. 2.7. (a) Solution of HHT (2.56) for the following input parameters: $v = 1$ nm/fs, $\tau = 1$ fs, $\omega\tau = 2$. (b) Solution of HHT equation for the following input parameters: $v = 1$ nm/fs, $\tau = 1$ fs, $\omega\tau = 0.45$.

Considering formulas (2.63) and (2.64), we observe that the thermal wave with velocity v_2 is very quickly attenuated in time. It occurs that when $\omega^{-1} > 2\tau$, the scatterings of the heat carriers diminish the thermal wave.

It is interesting to observe that in the limit of a very short relaxation time, i.e., when $\tau \to 0$, $v_2 \to \infty$, because for $\tau \to 0$, (2.56) is the Fourier parabolic equation.

It can be concluded that for $\omega^{-1} > 2\tau$, the Fourier equation is the relevant equation for the description of the thermal phenomena in metals. For $\omega^{-1} < 2\tau$, the scatterings are slower than in the preceding case and attenuation of the thermal wave is weaker. In that case, $\tau \neq 0$ and v_1 is always finite:

$$v_1 = v \left(1 - \left(\frac{1}{2\tau\omega} \right)^2 \right)^{1/2} < v. \qquad (2.65)$$

For $\tau \to \infty$, i.e., for very rare scatterings $v_1 \to v$, and (2.56) is a nearly free thermal wave equation. For τ finite, the $v_1 < v$ and thermal wave propagates in the medium with smaller velocity than the velocity of the initial thermal wave.

Considering the formula (2.62), one can define the change of the phase of the initial thermal wave β, i.e.,

$$\tan[\beta] = \frac{a}{w_1} = \frac{1}{2\tau\omega} \frac{1}{\sqrt{1 - \frac{1}{4\tau^2\omega^2}}}, \qquad 2\tau\omega > 1. \qquad (2.66)$$

We conclude that the scatterings produce the change of the phase of the initial thermal wave. For $\tau \to \infty$ (very rare scatterings), $\tan[\beta] = 0$.

In Figs. 2.7(a) and (b), the solutions of the (2.56) for the following input parameters are presented. In Fig. 2.7(a), $v = 1$ nm/fs [2.10], $\tau = 1$ fs, $\omega\tau = 2$, and the solution of (2.56), formula (2.62) represent the damped thermal wave that propagates with velocity $v_1 = 0.97v$. Figure 2.7(b) represents the solution of (2.56), formula (2.63) for the following input parameters: $v = 1$ nm/fs and $\omega\tau = 0.45$. In that case, the thermal wave is very quickly attenuated, and the solutions of (2.56), formula (2.63) represent the diffusion of the initial thermal wave.

2.2.4 Physics at the Attosecond Frontier

As every photographer knows, a flash of light can stop the action. Just as a fast flash lamp can freeze the image of a bullet in midflight, so short laser pulses can be used to probe fast molecular motion. It is no surprise then that laser scientists have been pushing for ever shorter pulses of light in order to follow ever-more rapid processes. More than a generation ago, the subpicosecond pulses were at the frontier of the laser possibility. Currently, table-top femtosecond laser pulses are in labs worldwide.

Femtosecond pulses led to femtochemistry the experimental study of fast chemical reactions and molecular dynamics. Even the fastest molecular vibrations appear completely still when probed with a pulse lasting a few femtoseconds. Now we are entering the era of attosecond pulses (1 attosecond is 10^{-18} s). M. Hentschel et al. [2.33] describe the generation and use of pulses lasting 650 attoseconds, in what might be down of attophysics – the study of dynamics on time scales fast enough to follow electronic motion within atoms.

What sort of experiments are possible with attosecond pulses? At these time scales, chemistry is essentially frozen in time, so the only dynamics to be studied are those of electrons, as they are much lighter and faster than nuclei. Most femtochemistry measurements are based on the pump–probe method: one pulse sets the system into motion and a second one probes it after a controlled delay. But this approach is not possible with attosecond laser pulses: they are just too weak. Moreover, manipulating such pump and probe pulses is harder to do because their wavelength is in the X-range, for which traditional optical tools (such as lenses and beamsplitters) do not exist.

The work of M. Hentschal et al. hints at the likely direction that attophysics will take in the near future. They achieve attosecond measurements by carefully timing the delay between an attosecond pulse and an optical pulse.

References

2.1. M. Kac in Magnolia Petroleum Co.: *Lectures in Pure and Applied Science*, No 2, (1956).

2.2. R. Hersh: Part. Diff. Eq. and Related Topics. In: *Lecture Notes in Math*, No 446 (Springer-Verlag 1975) pp 283–300

2.3. D. Jou, J. Casas-Vázques, G. Lebon: *Extended Irreversible Thermodynamics* (Springer Berlin 1993)

2.4. A.V. Luikov: *Analitical Heat Diffusion Theory* (Academic Press New York 1968)

2.5. Bo Martin Bibley, M. Sorensen: Finance Stochast. **1**, 25 (1997)

2.6.   A. Di Prisco, L. Herrera, M. Esculpi: Class. Quant. Grav. **13**, 1053 (1996)

2.7.   M. Kozlowski: Nucl. Phys. **A492**, 285 (1989)

2.8.   G.L. Eesley: Phys. Rev. Lett. **51**, 2140 (1983)

2.9.   S.D. Brorson, J.G. Fujimoto, E.P. Ippen: Phys. Rev. Lett. **59**, 1962 (1987)

2.10. H.E. Elsayed-Ali et al.: Phys. Rev. B **43**, 4488 (1991)

2.11. T. Juhasz et al.: Phys. Rev. B **45**, 13819 (1992)

2.12. W.S. Fann et al.: Phys. Rev. Lett. **68**, 2834 (1992)

2.13. W.S. Fann et al.: Phys. Rev. B **46**, 13592 (1992)

2.14. R.H.M. Groeneveld: Femtosecond Spectroscopy on Electrons and Phonons in
       Noble Metals. Ph. D. Thesis, Van der Waals-Zeeman Laboratory, University
       of Amsterdam (1992)

2.15. P.B. Corkum, F. Brunel, N.K. Sherman: Phys. Rev. Lett. **61**, 2886 (1988)

2.16. B.M. Clemens, G.L. Eesley, A.C. Paddock: Phys. Rev. B **37**, 1085 (1988)

2.17. J. Marciak-Kozlowska: Int. J. Thermophys. **14**, 593 (1993)

2.18. J. Marciak-Kozlowska: J. Phys. Ch. Solids **55**, 721 (1994)

2.19. J. Marciak-Kozlowska: Lasers Eng. **4**, 57 (1995)

2.20. J. Marciak-Kozlowska: Int. J. Thermophys. **16**, 1989 (1995)

2.21. Ch. Kittel, H. Kroemer: *Thermal Physics* (W. H. Freeman and Company San
       Francisco 1980)

2.22. D. Pines P. Nozieres: *The Theory of Quantum Liquids* (Benjamin New
       York 1966)

2.23. S.J. Anisimov, B.L. Kapoliovich, T.L. Pereleman: Sov. Phys. JETP **39**, 375
       (1975)

2.24. W. F. Królikowski, W.E. Spicer: Phys. Rev. B **1**, 478 (1970)

2.25. W. Nernst: *Die Theoretischen Grundlagen des Wärmestatzes* (Knapp
       Halle 1917)

2.26. R.B. Dingle: Proc. Phys. Soc. London **A65**, 374 (1952)

2.27. J.C. Ward, J. Wilks: Philos. Mag. **42**, 314 (1951)

2.28. F. London: *Superfluids* vol. **II** (J. Wiley and Sons New York 1954) p 101

2.29. J.C. Ward, J. Wilks: Philos. Mag. **43**, 48 (1952)

2.30. C.C. Ackerman, R.A. Guyer: Ann. Phys. **50**, 128 (1968)

2.31. Ya. B. Zeldovich, A.S. Kompaneets: *Collection Dedicated to the Seventieth
       Birthday of Academician A. F. Joffe* (Izd. Akad. Nauk SSR, 1959) p 61

2.32. C. De Witt-Morette, See Kit Fong: Phys. Rev. Lett. **19**, 2201 (1989)

2.33. M. Hentschel et al.: Nature **414**, 509 (2001)

Causal Thermal Phenomena, Quantal Description

3.1 Discretization of the Thermal Excitation in High Excited Matter

3.1.1 Quantum Heat Transport Equation (QHT)

There is an impressive amount of literature on hyperbolic heat transport in matter [3.1]–[3.5]. In Chapter 2, we developed the new hyperbolic heat transport equation, which generalizes the Fourier heat transport equation for rapid thermal processes. The hyperbolic heat transport equation (HHT) for the fermionic system has be written in the form (2.25)

$$\frac{1}{\left(\frac{1}{3}v_F^2\right)} \frac{\partial^2 T}{\partial t^2} + \frac{1}{\tau\left(\frac{1}{3}v_F^2\right)} \frac{\partial T}{\partial t} = \nabla^2 T, \tag{3.1}$$

where T denotes the temperature, τ the relaxation time for the thermal disturbance of the fermionic system, and v_F is the Fermi velocity.

In what follows, we develop the new formulation of the HHT, considering the details of the two fermionic systems: electron gas in metals and the nucleon gas.

For the electron gas in metals, the Fermi energy has the form

$$E_F^e = (3\pi)^2 \frac{n^{\frac{2}{3}}\hbar^2}{2m_e}, \tag{3.2}$$

where n denotes the density and m_e electron mass. Considering that

$$n^{-\frac{1}{3}} \sim a_B \sim \frac{\hbar^2}{me^2}, \tag{3.3}$$

and a_B = Bohr radius, one obtains

$$E_F^e \sim \frac{n^{\frac{2}{3}} \hbar^2}{m_e} \sim \frac{\hbar^2}{ma^2} \sim \alpha^2 m_e c^2 , \tag{3.4}$$

where c = light velocity, and $\alpha = 1/137$ is the fine-structure constant for electromagnetic interaction. For the Fermi momentum p_F, we have

$$p_F^e \sim \frac{\hbar}{a_B} \sim \alpha m_e c , \tag{3.5}$$

and, for Fermi velocity v_F,

$$v_F^e \sim \frac{p_F}{m_e} \sim \alpha c . \tag{3.6}$$

Formula (3.6) gives the theoretical background for the result presented in Chapter 2. Comparing formulas (2.48) and (3.6), it occurs that $b = \alpha$. Considering formula (3.6), Eq. (2.49) can be written as

$$\frac{1}{c^2} \frac{\partial^2 T}{\partial t^2} + \frac{1}{c^2 \tau} \frac{\partial T}{\partial t} = \frac{\alpha^2}{3} \nabla^2 T . \tag{3.7}$$

As seen from (3.7), the HHT equation is a relativistic equation, because it takes into account the finite velocity of light.

For the nucleon gas, Fermi energy equals

$$E_F^N = \frac{(9\pi)^{\frac{2}{3}} \hbar^2}{8mr_0^2} , \tag{3.8}$$

where m denotes the nucleon mass and r_0, which describes the range of strong interaction, is given by

$$r_0 = \frac{\hbar}{m_\pi c} , \tag{3.9}$$

wherein m_π is the pion mass. From formula (3.9), one obtains for the nucleon Fermi energy

$$E_F^N \sim \left(\frac{m_\pi}{m}\right)^2 mc^2 . \tag{3.10}$$

In analogy to Eq. (3.4), formula (3.10) can be written as

$$E_F^N \sim \alpha_s^2 mc^2 , \tag{3.11}$$

where $\alpha_s = \frac{m_\pi}{m} \cong 0.15$ is the fine-structure constant for strong interactions. Analogously, we obtain the nucleon Fermi momentum

$$p_F^N \sim \frac{\hbar}{r_0} \sim \alpha_s mc \tag{3.12}$$

and the nucleon Fermi velocity

$$v_F^N = \frac{pF}{m} \sim \alpha_s c,$$
(3.13)

and HHT for nucleon gas can be written as

$$\frac{1}{c^2} \frac{\partial^2 T}{\partial t^2} + \frac{1}{c^2 \tau} \frac{\partial T}{\partial t} = \frac{\alpha_s^2}{3} \nabla^2 T.$$
(3.14)

In the following, the procedure for the discretization of temperature $T(r,t)$ in hot fermion gas will be developed. First of all, we introduce the reduced de Broglie wavelength

$$\lambdabar_B^e = \frac{\hbar}{m_e v_h^e}, \qquad v_h^e = \frac{1}{\sqrt{3}} \alpha c,$$

$$\lambdabar_B^N = \frac{\hbar}{m v_h^N}, \qquad v_h^N = \frac{1}{\sqrt{3}} \alpha_s c,$$
(3.15)

and the mean free paths λ^e and λ^N

$$\lambda^e = v_h^e \tau^e, \qquad \lambda^N = v_h^N \tau^N.$$
(3.16)

In view of formulas (3.15) and (3.16), we obtain the HHC for electron and nucleon gases

$$\frac{\lambdabar_B^e}{v_h^e} \frac{\partial^2 T^e}{\partial t^2} + \frac{\lambdabar_B^e}{\lambda^e} \frac{\partial T}{\partial t} = \frac{\hbar}{m_e} \nabla^2 T^e,$$
(3.17)

$$\frac{\lambdabar_B^N}{v_h^N} \frac{\partial^2 T^N}{\partial t^2} + \frac{\lambdabar_B^N}{\lambda^N} \frac{\partial T^N}{\partial t} = \frac{\hbar}{m} \nabla^2 T^N.$$
(3.18)

Equations (3.17) and (3.18) are the hyperbolic partial differential equations that are the master equations for heat propagation in Fermi electron and nucleon gases. In the following, we will study the quantum limit of heat transport in the fermionic systems. We define *the quantum heat transport limit* as follows:

$$\lambda^e = \lambdabar_B^e, \qquad \lambda^N = \lambdabar_B^N.$$
(3.19)

In that case, Eqs. (3.17) and (3.18) have the form

$$\tau^e \frac{\partial^2 T^e}{\partial t^2} + \frac{\partial T^e}{\partial t} = \frac{\hbar}{m_e} \nabla^2 T^e,$$
(3.20)

$$\tau^N \frac{\partial^2 T^N}{\partial t^2} + \frac{\partial T^N}{\partial t} = \frac{\hbar}{m} \nabla^2 T^N,$$
(3.21)

where

$$\tau^e = \frac{\hbar}{m_e (v_h^e)^2}, \qquad \tau^N = \frac{\hbar}{m (v_h^N)^2}.$$
(3.22)

Equations (3.20) and (3.21) define the master equation for quantum heat transport (QHT). Having the relaxation times τ^e and τ^N, one can define the "pulsations" ω_h^e and ω_h^N

$$\omega_h^e = (\tau^e)^{-1}, \qquad \omega_h^N = (\tau^N)^{-1}, \tag{3.23}$$

or

$$\omega_h^e = \frac{m_e(v_h^e)^2}{\hbar}, \qquad \omega_h^N = \frac{m(v_h^N)^2}{\hbar},$$

i.e.,

$$\omega_h^e \hbar = m_e(v_h^e)^2 = \frac{m_e \alpha^2}{3} c^2,$$

$$\omega_h^N \hbar = m(v_h^N)^2 = \frac{m \alpha_s^2}{3} c^2. \tag{3.24}$$

The formulas (3.24) define the Planck-Einstein relation for heat quanta E_h^e and E_h^N

$$E_h^e = \omega_h^e \hbar = m_e(v_h^e)^2,$$

$$E_h^N = \omega_h^N = m_N(v_h^N)^2. \tag{3.25}$$

The heat quantum with energy $E_h = \hbar\omega$ can be named the *heaton*, in complete analogy to the *phonon, magnon, roton*, etc. For τ^e, $\tau^N \to 0$, Eqs. (3.20) and (3.24) are the Fourier equations with quantum diffusion coefficients D^e and D^N

$$\frac{\partial T^e}{\partial t} = D^e \nabla^2 T^e, \qquad D^e = \frac{\hbar}{m_e}, \tag{3.26}$$

$$\frac{\partial T^N}{\partial t} = D^N \nabla^2 T^N, \qquad D^N = \frac{\hbar}{m}. \tag{3.27}$$

The quantum diffusion coefficients D^e and D^N were introduced for the first time by E. Nelson [3.6].

For finite τ^e and τ^N, for $\Delta t < \tau^e$, $\Delta t < \tau^N$, Eqs. (3.20) and (3.21) can be written as

$$\frac{1}{(v_h^e)^2} \frac{\partial^2 T^e}{\partial t^2} = \nabla^2 T^e, \tag{3.28}$$

$$\frac{1}{(v_h^N)^2} \frac{\partial^2 T^N}{\partial t^2} = \nabla^2 T^N. \tag{3.29}$$

Equations (3.28) and (3.29) are the wave equations for quantum heat transport (QHT). For $\Delta t > \tau$, one obtains the Fourier equations (3.26) and (3.27).

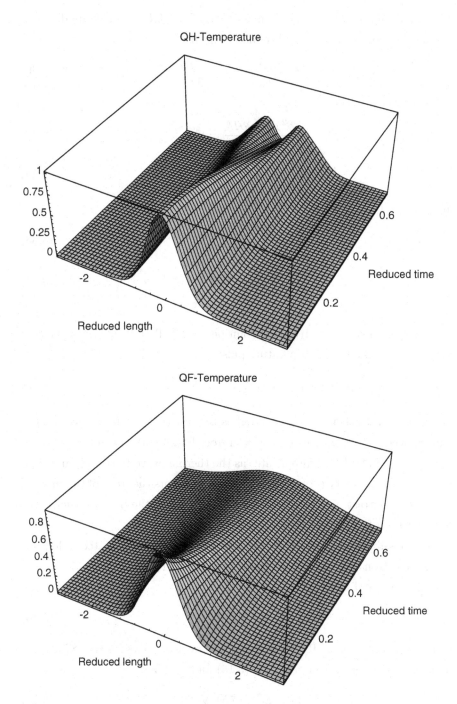

Fig. 3.1. (a) The numerical solution of the QHT (3.31) for the initial Gaussian temperature profile (3.35). (b) The numerical solution of the QFT (3.33) for the initial Gaussian temperature profile (3.35).

In what follows, the dimensionless form of the QHT will be used. Introducing the reduced time t' and reduced length x',

$$t' = \frac{t}{\tau}, \qquad x' = \frac{x}{v_h \tau}, \tag{3.30}$$

one obtains, for QHT,

$$\frac{\partial^2 T^e}{\partial t^2} + \frac{\partial T^e}{\partial t} = \nabla^2 T^2, \tag{3.31}$$

$$\frac{\partial^2 T^N}{\partial t^2} + \frac{\partial T^N}{\partial t} = \nabla^2 T^N \tag{3.32}$$

and, for QFT,

$$\frac{\partial T^e}{\partial t} = \nabla^2 T^e, \tag{3.33}$$

$$\frac{\partial T^N}{\partial t} = \nabla^2 T^N. \tag{3.34}$$

In Figs. 3.1(a) and 3.1(b), the solutions of QHT and QFT, respectively, for the initial Gaussian temperature pulse

$$T(0, x') = \exp(-3x'^2), \qquad \frac{\partial T(t'x')}{\partial t'}\bigg|_{t'=0} = 0 \tag{3.35}$$

are presented. Isotherms are presented as the functions of the reduced length and reduced time (3.30). As can be seen from Fig. 3.1(a), the initial Gaussian temperature pulse (3.35) propagates as the thermal wave to the left and right with the same velocity ("two arms" picture represents the front of the thermal wave). By contrast, in the case of QFT (Fig. 3.1(b)), the heat transfer is the heat diffusion from the thermal Gaussian source.

The possible interpretation of the heaton energies can be stated as follows. For an electron gas, we obtain from formulas (3.15) and (3.25), for $m_e = 0.51 \, \text{MeV}/c^2$, and $v_h = (1/\sqrt{3})\alpha c$,

$$E_h^e = 9 \, \text{eV}, \tag{3.36}$$

which is of the order the Rydberg energy. For nucleon gases ($m = 938 \, \text{MeV}/c^2$, $\alpha_s = 0.15$) one finds, from formulas (3.15) and (3.25),

$$E_h^N \sim 7 \, \text{MeV} \tag{3.37}$$

i.e., the average binding energy of the nucleon in the nucleus ("boiling" temperature for the nucleus).

When the ordinary matter (on the atomic level) or nuclear matter (on the nucleus level) is excited with short temperature pulses ($\Delta t \sim \tau$), the response of the matter is discrete. The matter absorbs the thermal energy in the form of the quanta E_h^e or E_h^N.

It is quite natural to pursue the study of the thermal excitation to the subnucleon level, i.e., quark matter. In the following, we generalize the QHT equation (3.7) for quark gas in the form

$$\frac{1}{c^2}\frac{\partial^2 T^q}{\partial t^2} + \frac{1}{c^2\tau}\frac{\partial T^q}{\partial t} = \frac{(\alpha_s^q)^2}{3}\nabla^2 T^q, \tag{3.38}$$

with α_s^q denoting the fine-structure constant for strong quark-quark interaction, v_h^q the thermal velocity

$$v_h^q = \frac{1}{\sqrt{3}}\alpha_s^q c, \tag{3.39}$$

and τ is relaxation time for quark gas.

Analogously to electron and nucleon gases, we obtain for quark heaton

$$E_h^q = \frac{m_q}{3}(\alpha_s^q)^2 c^2, \tag{3.40}$$

where m_q denotes the mass of the average quark mass. For a quark gas, the average quark mass can be calculated according to formula, [3.7]

$$m_q = \frac{1}{3}(m_u + m_d + m_s)$$

$$= \frac{1}{3}(350 + 350 + 550)\,\text{MeV} = 417\,\text{MeV}, \tag{3.41}$$

where m_u, m_d, m_s denotes the mass of the up, down, and strange quark, respectively. For the calculation of the α_s^q, we consider the decays of the baryon resonances. For strong decay of the $\Sigma^0(1385\,\text{MeV})$ resonance

$$K^- + p \rightarrow \Sigma^0(1385\,\text{MeV}) \rightarrow \Lambda + \pi^0,$$

the width $\Gamma \sim 36\,\text{MeV}$ and lifetime

$$\tau_s = \frac{\hbar}{\Gamma} \sim 10^{-23}\,\text{s}.$$

For electromagnetic decay,

$$\Sigma^0(1192\,\text{MeV}) \rightarrow \Lambda + \gamma,$$

$\tau_e \sim 10^{-19}\,\text{s}$. Considering that

$$\left(\frac{\alpha_s^q}{\alpha}\right) \sim \left(\frac{T_e}{T_s}\right)^{1/2} \sim 100 \, ,$$

one obtains for α_s^q the value

$$\alpha_s^q \sim 1 \, . \tag{3.42}$$

Substituting formulas (3.41) and (3.42) into (3.40), one finds

$$E_h^q \sim 139 \, \text{MeV} \sim m_\pi \, , \tag{3.43}$$

where m_π denotes the π-meson mass. It occurs that where one attempts to "melt" the nucleon in order to obtain the free quark gas, the energy of the heaton is equal to the π-meson mass (which consists of two quarks). It is the simple presentation of *quark confinement*.

The contemporary results of the investigation of havy-ion interactions seems to support the above considerations. With the advent of ultra-relativistic heavy-ion collisions in the laboratory, at CERN and Brookhaven a new inter-disciplinary field has emerged from the traditional domains of particle physics and nuclear physics. In combining methods and concepts from both areas, the study of heavy-ion interactions at very high energies, a new and orig-inal approach in investigating the properties of matter and its interactions. Combining the elementary-interactions aspect of high-energy physics with the macroscopic-matter aspect of nuclear physics, the subject of heavy-ion colli-sions is the study of *bulk matter* consisting of strongly interacting particles (hadrons). Thermodynamics would be an ideal language to be used in this field, so that complex multiparticle states can be described in terms of a few macroscopic variables: temperature, density, etc.

3.1.2 Brownian Representation of Quantum Heat Transport in Attosecond Domain

The advent of the ultrashort duration laser pulses, particularly those in at-tosecond (10^{-18} s) domain [3.8], open new experimental possibilities in the study of the details of the quantum heat transport, e.g. the quantum path of the heat carriers.

The quantum heat transport equation (QHT), formulas (3.20), (3.21) de-scribe the quantum limit of heat transport. From a mathematical point of view, the equations (3.20), (3.21) are the telegraph type equation. In pa-per [3.9], it was shown that the hyperbolic diffusion equation (telegraph equa-tion) can be obtained within the frame of the correlated random walk (CRW) theory of the Brownian motion. As was shown in paper [3.9], the average displacement of the Brownian particle is described by the formula

$$< x^2 > = \frac{2\hbar\tau}{m_e} \left[\frac{t}{\tau} - \left(1 - e^{-\frac{t}{\tau}} \right) \right] . \tag{3.44}$$

In deriving Eq. (3.44), the diffusion coefficient $D = \hbar/m$ was used, according to formula (3.26). Equation (3.44) has important consequences for understanding the quantum heat transport. Can we define the trajectory of the *heatons*? To find the answer, let us discuss the two time limits of Eq. (3.44). First of all, let us assume $t \gg \tau$. Then, from formula, one obtains

$$< x^2 > \cong \frac{2\hbar\tau}{m_e} \left(\frac{t}{\tau} - 1 \right) = \frac{2\hbar}{m_e} t . \tag{3.45}$$

Equation (3.45) describes the quantum diffusion of the *heatons* with quantum diffusion coefficient $D = \hbar/m$, i.e.,

$$< x^2 > = 2Dt . \tag{3.46}$$

It is interesting to observe that Eq. (3.46) can be interpreted as describing the random walk of a Brownian particles (*heatons*) and that the *heaton* path has fractal dimension $d_f = 2$, because the "mass" segment with duration t is related to the "radius" x by the relation (3.46) [3.10]

$$t \sim x^{d_f} . \tag{3.47}$$

The fractal dimension of the quantum path was investigated by Abbot and Wise [3.11]. They showed that the observed path of a particle in quantum mechanics is a fractal curve with the Hausdorff dimension two.

For $t \sim \tau$, one obtains from Eq. (3.44)

$$< x^2 > \cong \frac{2\hbar\tau}{m} \left(\frac{t}{\tau} - \left(1 - 1 + \frac{t}{\tau} - \frac{t^2}{2\tau^2} \right) \right) = \frac{\hbar}{m\tau} t^2 , \tag{3.48}$$

which in view of formula (3.22) gives

$$< x^2 > = v_h^2 t^2 . \tag{3.49}$$

Equation (3.49) describes the free motion of a *heaton* with velocity $v_h = \frac{1}{\sqrt{3}}\alpha c$ and considering Eq. (3.47), the path of quantum particle has a fractal dimension $d_f = 1$, i.e. the straight line. Now, we show that the analogy between quantum paths and random walks of quantum particles (*heaton*) can be extended. As was shown in Section 2.1, for $\Delta t < \tau$ the QHT has the form of the quantum wave equation

$$\tau \frac{\partial^2 T}{\partial t^2} = \frac{\hbar}{m_e} \nabla^2 T \tag{3.50}$$

i.e.

$$\frac{1}{v_h^2} \frac{\partial^2 T}{\partial t^2} = \nabla^2 T. \tag{3.51}$$

The maximum value of the thermal wave velocity can be equal to c – the velocity of light

$$v_h = c. \tag{3.52}$$

In that case, the relaxation time is described by formula

$$\tau_r = \frac{\hbar}{m v_h^2} \quad\longrightarrow\quad \frac{\hbar}{m c^2}. \tag{3.53}$$

The relaxation time τ_r, for $v_h = c$ is given by

$$\tau_r = \frac{\Lambda}{c}, \tag{3.54}$$

where Λ denotes the reduced Compton wavelength for electron,

$$\Lambda = \frac{\hbar}{m_e c}. \tag{3.55}$$

Having established the expression for relaxation time τ_r, the "pulsation" ω_r can be defined as

$$\omega_r = \tau_r^{-1} = \frac{m c^2}{\hbar} \tag{3.56}$$

i.e.

$$E_h^e = \omega_r \hbar = m_e c^2. \tag{3.57}$$

Considering that $m_e = 0.511\,\text{MeV}/c^2$, the *heaton* internal energy can be calculated

$$E_h^e = 0.511\,\text{MeV}, \tag{3.58}$$

and diffusion coefficient $D = \hbar/m$ can be written as

$$D = \frac{\hbar}{m} = \frac{\hbar}{m c} c = \Lambda c. \tag{3.59}$$

It is well-known from quantum electrodynamics that quantum void fluctuations create and destroy virtual electron-positon pairs. These virtual electron-positon pairs have a characteristic lifetime of the order Λ/c. This is the typical time scale over which collisions occur between the *heaton* and the virtual electron-positon pairs. It is clear that the Compton wavelength Λ can be identified as the new mean free path because it is the typical distance covered by the virtual pair before its annihilation. The Eq. (3.59) for diffusion coefficient relates the D to Λ in the same manner as in kinetic theory of gases, i.e.

$$D = \lambda_e v \qquad \longrightarrow \qquad D = \Lambda c. \qquad (3.60)$$

For the relativistic regime, the average displacement $< x^2 >$ has the form of equations (3.46) and (3.48)

$$< x^2 >= 2Dt \qquad (3.61)$$

for $t \gg \tau_r$ and,

$$< x^2 >= c^2 t^2$$

for $t < \tau_r$.

As in nonrelativistic regime, for $t \gg \tau_r$, the quantum path has the fractal dimension $d_f = 2$, and for $t < \tau_r$ the quantum path is the straight line but now with velocity $v_h = c$.

The advent of ultrashort duration pulses, particularly those in attosecond domain, has opened up new experimental possibilities in the study of structure of the matter in sub-nanometer scale. Conditions in this new field of investigation are markedly different from those for longer pulse duration. The distinction being that at longer pulse duration (~ 1 ps), excited particles and their surroundings have had sufficient time to approach thermal equilibrium. For temporal resolution ~ 1 fs, it is possible to resolve the dynamics of the nonequilibrium excited carrier. The time resolution of the order 1 attosecond (10^{-18} s) offers the possibilities of observing the path of the quantum heat carriers - *heatons*.

For a temporal resolution Δt of the order the relaxation time $\tau \sim 10^{-17}$ s, the erratic Brownian motion of the individual *heatons* for $\Delta t \gg \tau$ can be observed. For $\Delta t < \tau$, the *heatons* move along straight lines. Lasers with attosecond laser pulses open up quite new possibilities for studying these discrete thermal phenomena.

For the contemporary laser technology, the time resolution Δt of the order relaxation time $\tau_r \sim 10^{-21}$ s (formula (3.58)) is out of technological possibilities. Nevertheless, it is important to take into account the fact that for $\Delta t \leq 10^{-21}$ s, the *heatons* move with the velocity of the light! It is worthwhile to realize that there exist models of elementary particles in which it is assumed that the electron propagates with the speed of light with certain chirality, except that at random times it flips both the direction of propagation (by 180°) and handedness and, the rate of such flips is precisely the mass m (in units $\hbar = c = 1$) [3.12]–[3.15]. In the frame of the model developed in the current paper, the relaxation time for the interaction of *heatons* with voids is described by the formula (3.58) as $\tau = \hbar/mc^2$.

3.1.3 The Fundamental Solution of the Quantum Heat Transport Equation

In what follows, we will describe the one-dimensional hyperbolic heat transfer with the help of (3.20) and (3.21). To that aim, let us consider the following transformation of the temperature field $T(x,t)$

$$T(x,t) = e^{-\frac{t}{2\tau^{e,N}}} U^{e,N}(x,t). \tag{3.62}$$

When applying (3.62) to (3.20) and (3.21), one obtains the new QHT

$$\frac{\partial^2 U^{e,N}(x,t)}{\partial t^2} - (v_h^{e,N})^2 \frac{\partial^2 U^{e,N}(x,t)}{\partial x^2}$$

$$-\frac{1}{4(\tau^{e,N})^2} U^{e,N}(x,t) = 0. \tag{3.63}$$

For the Cauchy boundary conditions

$$U^{e,N}(x,0) = f^{e,N}(x), \qquad \left.\frac{\partial U^{e,N}(x,t)}{\partial t}\right|_{t=0} = g^{e,N}(x), \tag{3.64}$$

the solution of (3.63) has the form [3.16]

$$U^{e,N}(x,t) = \frac{f^{e,N}(x - v_h^{e,N}t) + f^{e,N}(x + v_h^{e,N}t)}{2} \tag{3.65}$$

$$+ \frac{1}{2v_h^{e,N}} \int_{x-v_h^{e,N}t}^{x+v_h^{e,N}t} g(\zeta) I_0 \left[\frac{1}{2\tau^{e,N}v_h^{e,N}}\sqrt{(v_h^{e,N}t)^2 - (x-\zeta)^2}\right] d\zeta$$

$$+ \frac{t}{2\tau^{e,N}} \int_{x-v_h^{e,N}t}^{x+v_h^{e,N}t} f(\zeta) \frac{I_1\left[\frac{1}{2\tau^{e,N}v_h^{e,N}}\sqrt{(v_h^{e,N}t)^2 - (x-\zeta)^2}\right]}{\sqrt{(v_h^{e,N}t)^2 - (x-\zeta)^2}} d\zeta,$$

and the solutions of (3.20), (3.21) are

$$T^{e,N}(x,t) = e^{-t/2\tau^{e,N}} U^{e,N}(x,t). \tag{3.66}$$

For $\tau^{e,N} \to \infty$ (ballistic quantum heat transport), (3.63) can be written as

$$\frac{\partial^2 U^{e,N}}{\partial t^2} - (v_h^{e,N})^2 \frac{\partial^2 U^{e,N}}{\partial x^2} = 0 \tag{3.67}$$

and for Cauchy boundary conditions (3.64), the wave equation (3.67) has the solution [3.16]

$$U^{e,N}(x,t) = \frac{1}{2} \left[f^{e,N}(x + v_h^{e,N}t) + f(x - v_h^{e,N}t) \right]$$

$$+ \frac{1}{2v_h^{e,N}} \int_{x-v_h^{e,N}t}^{x+v_h^{e,N}t} g(\zeta) d\zeta . \tag{3.68}$$

In that case the solutions of (3.20) and (3.21) are

$$T^{e,N}(x,t) = U^{e,N}(x,t). \tag{3.69}$$

Both solutions (3.66) and (3.69) exhibit the domains of dependence and influence on the hyperbolic equations. These domains, which characterize the maximum speed at which disturbances or signals travel, are determined by the principal parts of the given equations (i.e. the second derivative terms) and do not depend on the lower order terms. These results show that the QHT and the wave equation have identical domains of dependence and influence.

Now, let us consider the static field limit $U^{e,N}(x,t) \rightarrow V^{e,N}(x)$ of Eq. (3.63)

$$\frac{\partial^2 V^e(x)}{\partial x^2} = -\frac{1}{4} \left(\frac{m_e v_h^e}{\hbar} \right)^2 V^e(x),$$

$$\frac{\partial^2 V^N(x)}{\partial x^2} = -\frac{1}{4} \left(\frac{m v_h^N}{\hbar} \right)^2 V^N(x). \tag{3.70}$$

If we require spherically symmetric solutions of (3.70), i.e. one that solely depends upon $|r|$ (in three dimensions), then we can write

$$\nabla^2 V^e(r) = -\frac{1}{4} \left(\frac{m_e v_h^e}{\hbar} \right)^2 V^e(r),$$

$$\nabla^2 V^N(r) = -\frac{1}{4} \left(\frac{m v_h^N}{\hbar} \right)^2 V^N(r). \tag{3.71}$$

The spherically symmetric solutions of (3.71) are

$$V^e(r) = -\frac{g^e}{r} e^{-\frac{r}{R^e}},$$

$$V^N(r) = -\frac{g^N}{r} e^{-\frac{r}{R^N}}. \tag{3.72}$$

Table 3.1. The Ranges and Heaton Energies for Electromagnetic and Strong Interactions

Fermions	R^e, R^N (m)	e^e, e^N (eV)
Electrons	10^{-10}	~ 10
Nucleons	10^{-15}	$\sim 10^7$

where $g^{e,N}$ denotes the coupling constants ("charges") for electromagnetic and strong interactions, respectively. The parameters R^e and R^N are the ranges of the interactions. Following formula (3.72), one obtains

$$R^e = \frac{2\hbar}{m_e v_h^e}, \qquad R^N = \frac{2\hbar}{m_N v_h^N}. \qquad (3.73)$$

For electromagnetic interactions, the range R^e can be written in the form

$$R^e \cong \frac{2\hbar}{m_e v_F^e} \sim \frac{2\hbar}{P_F^e} \sim \frac{2}{k_F^e}. \qquad (3.74)$$

where k_F^e denotes the Fermi wave vector for the electrons. Substituting formula (3.74) to (3.72), the potential $V^e(r)$ can be written as

$$V^e(r) = -\frac{g^e}{r} e^{-\frac{rk_F}{2}}. \qquad (3.75)$$

Formula (3.75) is the well-known equation for Debye-Hückl Coulomb potential with screening [3.17, 3.18], where g^e is equal $\alpha\hbar c$.

For strong interactions, potential $V^N(r)$ has the form of the Yukawa nucleon-nucleon potential [3.19]

$$V^N(r) = -\frac{g^N}{r} e^{-\frac{r}{R^N}}, \qquad g^N = \alpha_s\hbar c, \qquad (3.76)$$

with the range R^N, where

$$R^N = \frac{2\hbar}{mv_h^N} = \frac{2\hbar}{m_N \alpha_s c}, \qquad (3.77)$$

where $\alpha_s = m_\pi m_N^{-1}$ is the fine-structure constant for strong interactions. In that case, formula (3.77) can be written as

$$R^N = \frac{2\hbar}{m_\pi c}, \qquad (3.78)$$

i.e. R^N is of the order Compton wavelength for meson π, which defines the range of strong interactions. Following formulas (3.25), (3.73), (3.74), and (3.78), the numerical values for *heaton* energies e^N and ranges of interactions are calculated and presented in Table 3.1.

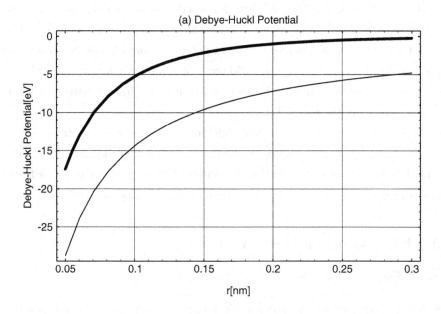

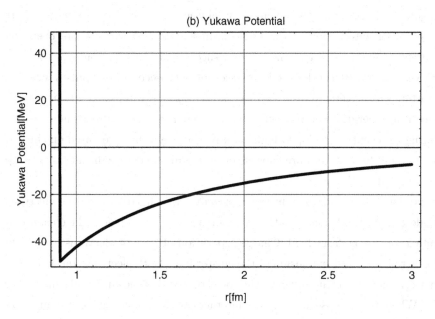

Fig. 3.2. (a) The Debye-Hückl potential, formula (3.75) (thick curve) and Coulomb potential (thin curve). **(b)** The Yukawa potential, formula (3.76), with hard-core term.

In Fig. 3.2(a), the Debye-Hückl potential, formula (3.75), is presented (thick curve). In the same figure, the Coulomb potential (thin curve) is also presented. The nucleon-nucleon Yukawa potential, formulas (3.76), (3.78), is presented in Fig. 3.2(b). As it is well-known [3.19], for $r < 0.8\,\text{fm}$ the nucleon-nucleon potential has the hard-core term. In Fig. 3.2(b), the hard-core term is represented by the vertical line.

The formulas (3.76) and (3.78), which describe the range of strong force, can be formulated in the spirit of Skyrme model [3.20]. Skyrme introduced a model of nucleons as distributions of pion fields. In our case, the strong force is mediated by meson π, which is a "part" of nucleon, i.e. $m_\pi \approx 0.15m$.

3.2 Klein-Gordon Thermal Equation

3.2.1 The Distortionless Quantum Thermal Waves

Efficient conversion of electromagnetic energy to particle energy is of fundamental importance in many areas of physics. The nature of intense, short pulse laser interactions with single atoms and solid targets has been the subject of extensive experimental and theoretical investigation over the past 15 years [3.21]. Recently, the interaction of femtosecond laser pulses with Xe clusters was investigated [3.22, 3.23] and strong X-ray emission and multi-keV electron generation were observed. Such experiments have become possible, owing to recently developed high peak power lasers that are based on chirped pulse amplification and are capable of producing focused light intensity of up to $10^{14} - 10^{19}\,\text{W/cm}^2$.

In intensely irradiated clusters, optically and collisionally ionized electrons undergo rapid collisional heating for short time (< 1 ps) before the cluster disintegrates in the laser field. Charge separation of the hot electrons inevitably leads to a very fast expansion of the cluster ions. Both electrons and ions ultimately reach a velocity given by the speed of sound of the cluster plasma [3.24].

When the intense laser pulse interacts with atomic clusters, ionization to very high charge states is observed [3.24]. The high Coulomb field certainly influences the thermal processes in clusters. In the chapter, the new QHT equation is formulated in which the external – not only thermal forces – are included. The solution of the new QHT for Cauchy boundary conditions will be derived. The condition for the distortionless propagations of the thermal wave will be formulated.

Now, we develop the generalized quantum heat transport equation, which includes the potential term. In this way, we use the analogy between the Schrödinger equation and quantum heat transport equations (3.31), (3.32). Let us consider, for the moment, the parabolic heat transport equation, i.e. the (3.33), (3.34), with the second derivative term omitted [3.9, 3.25]

$$\frac{\partial T}{\partial t} = \frac{\hbar}{m} \nabla^2 T \,. \tag{3.79}$$

When the real time $t \to \frac{it}{2}$ and $T \to \Psi$, (3.79) has the form of a free Schrödinger equation

$$i\hbar \frac{\partial \Psi}{\partial t} = -\frac{\hbar^2}{2m} \nabla^2 \Psi \,. \tag{3.80}$$

The complete Schrödinger equation has the form

$$i\hbar \frac{\partial \Psi}{\partial t} = -\frac{\hbar^2}{2m} \nabla^2 \Psi + V\Psi \,, \tag{3.81}$$

where V denotes the potential energy. When we go back to real time $t \to -2it$ and $\Psi \to T$, the new parabolic quantum heat transport is obtained

$$\frac{\partial T}{\partial t} = \frac{\hbar}{m} \nabla^2 T - \frac{2V}{\hbar} T \,. \tag{3.82}$$

Equation (3.82) describes the quantum heat transport for $\Delta t > \tau$. For heat transport initiated by ultrashort laser pulses, when $\Delta t \leq \tau$ one obtains the generalized quantum hyperbolic heat transport equation

$$\tau \frac{\partial^2 T}{\partial t^2} + \frac{\partial T}{\partial t} = \frac{\hbar}{m} \nabla^2 T - \frac{2V}{\hbar} T \,. \tag{3.83}$$

Considering that $\tau = \hbar/mv^2$ [3.9, 3.25], equation (3.83) can be written as follows:

$$\frac{1}{v^2} \frac{\partial^2 T}{\partial t^2} + \frac{m}{\hbar} \frac{\partial T}{\partial t} + \frac{2Vm}{\hbar^2} T = \nabla^2 T. \tag{3.84}$$

Equation (3.84) describes the heat flow when apart from the temperature gradient, the potential energy V operates.

In the following, we consider the one-dimensional heat transfer phenomena, i.e.

$$\frac{1}{v^2} \frac{\partial^2 T}{\partial t^2} + \frac{m}{\hbar} \frac{\partial T}{\partial t} + \frac{2Vm}{\hbar^2} T = \frac{\partial^2 T}{\partial x^2} \,. \tag{3.85}$$

For quantum heat transfer equation (3.85), we seek solution in the form

$$T(x,t) = e^{-\frac{t}{2\tau}} u(x,t) \,. \tag{3.86}$$

After substitution of (3.86) into (3.85), one obtains

$$\frac{1}{v^2}\frac{\partial^2 u}{\partial t^2} - \frac{\partial^2 u}{\partial x^2} + q\,u(x,t) = 0,\tag{3.87}$$

where

$$q = \frac{2Vm}{\hbar^2} - \left(\frac{mv}{2\hbar}\right)^2.\tag{3.88}$$

In the following, we will consider the constant potential energy $V = V_0$. The general solution of (3.87) for Cauchy boundary conditions,

$$u(x,0) = f(x), \qquad \left.\frac{\partial u(x,t)}{\partial t}\right|_{t=0} = F(x),\tag{3.89}$$

has the form [3.26]

$$u(x,t) = \frac{f(x - vt) + f(x + vt)}{2} + \frac{1}{2}\int_{x-vt}^{x+vt}\Phi(x,t,z)\mathrm{d}z,\tag{3.90}$$

where

$$\Phi(x,t,z) = \frac{1}{v}F(z)J_0\left(\frac{b}{v}\sqrt{(z-x)^2 - v^2 t^2}\right)\tag{3.91}$$

$$+bt f(z)\frac{J_0'(\frac{b}{v}\sqrt{(z-x)^2 - v^2 t^2})}{\sqrt{(z-x)^2 - v^2 t^2}},$$

$$b = \left(\frac{mv^2}{2\hbar}\right)^2 - \frac{2Vm}{\hbar^2}v^2\tag{3.92}$$

and $J_0(z)$ denotes the Bessel function of the first kind. Considering formulas (3.86), (3.87), and (3.88), the solution of (3.85) describes the propagation of the distorted thermal quantum waves with characteristic lines $x = \pm vt$. We can define the distortionless thermal wave as the wave that preserves the shape in the field of the potential energy V_0. The condition for conserving the shape can be formulated as

$$q = \frac{2Vm}{\hbar^2} - \left(\frac{mv}{2\hbar}\right)^2.\tag{3.93}$$

When (3.93) holds, (3.87) has the form

$$\frac{\partial^2 u(x,t)}{\partial t^2} = v^2\frac{\partial^2 u}{\partial x^2}.\tag{3.94}$$

Equation (3.94) is the quantum wave equation with the solution (for Cauchy boundary conditions (3.89))

$$u(x,t) = \frac{f(x - vt) + f(x + vt)}{2} + \frac{1}{2v}\int_{x-vt}^{x+vt}F(z)\mathrm{d}z.\tag{3.95}$$

It is quite interesting to observe that condition (3.93) has an analog in the classical theory of the electrical transmission line. In the context of the transmission of an electromagnetic field, the condition $q = 0$ describes the Heaviside distortionless line. Equation (3.93) – the distortionless condition – can be written as

$$V_0\tau \sim \hbar, \tag{3.96}$$

We can conclude that in the presence of the potential energy V_0, one can observe the undisturbed quantum thermal wave only when the Heisenberg uncertainty relation for thermal processes (3.96) is fulfilled.

The generalized quantum heat transport equation (GQHT) (3.85) leads to the generalized Schrödinger equation. After the substitution $t \to it/2$, $T \to \Psi$ in (3.85), one obtains the generalized Schrödinger equation (GSE)

$$i\hbar\frac{\partial\Psi}{\partial t} = -\frac{\hbar^2}{2m}\nabla^2\Psi + V\Psi - 2\tau\hbar\frac{\partial^2\Psi}{\partial t^2}. \tag{3.97}$$

Considering that $\tau = \hbar/mv^2 = \hbar/m\alpha^2c^2$ ($\alpha = 1/137$ is the fine-structure constant for electromagnetic interactions), (3.97) can be written as

$$i\hbar\frac{\partial\Psi}{\partial t} = -\frac{\hbar^2}{2m}\nabla^2\Psi + V\Psi - \frac{2\hbar^2}{m\alpha^2c^2}\frac{\partial^2\Psi}{\partial t^2}. \tag{3.98}$$

One can conclude that for time period $\Delta t < \hbar/m\alpha^2c^2 \sim 10^{-17}$ s, the description of quantum phenomena needs some revision. On the other hand, for $\Delta t > 10^{-17}$ in GSE, the second derivative term can be omitted, and as the result the SE is obtained, i.e.

$$i\hbar\frac{\partial\Psi}{\partial t} = -\frac{\hbar^2}{2m}\nabla^2\Psi + V\Psi. \tag{3.99}$$

It is quite interesting to observe that GSE was discussed also in papers [3.25, 3.27] in the context of the sub-quantal phenomena.

Concluding, a study of the interactions of the attosecond laser pulses with matter can shed light on the applicability of the SE to the study of the ultrashort sub-quantal phenomena. The structure of (3.87) depends on the sign of the parameter q. For quantum heat transport phenomena with electrons as the heat carriers, parameter q is the function of potential barrier height (V_0) and velocity v. Considering that velocity v equals [3.25]

$$v = \frac{1}{\sqrt{3}}\alpha c = 1.26\,\frac{\mathrm{nm}}{\mathrm{fs}}, \tag{3.100}$$

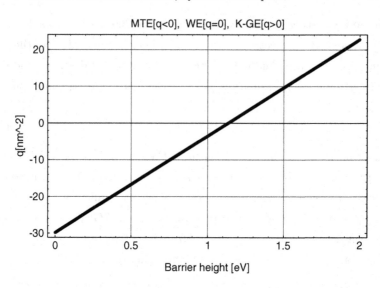

Fig. 3.3. Parameter q (formula (3.88)) as the function of the barrier height (eV).

parameter q can be calculated for typical barrier height $V_0 \geq 0$. In Fig. 3.3, the parameter q as the function of V_0 is calculated. For $q < 0$, i.e., when $V_0 < 1.125$ eV, (3.87) is the *modified telegraph equation* [3.14].

For Cauchy initial condition

$$u(x,o) = f(x), \qquad \frac{\partial u(x,o)}{\partial t} = g(x), \qquad (3.101)$$

the solution of Eq. (3.87) has the form

$$u(x,t) = \frac{f(x-vt) + f(x+vt)}{2} \qquad (3.102)$$

$$+ \frac{1}{2v} \int_{x-vt}^{x+vt} g(\zeta) I_0 \left[\sqrt{-q(v^2t^2 - (x-\zeta)^2)} \right] d\zeta$$

$$+ \frac{(v\sqrt{-q})t}{2} \int_{x-vt}^{x+vt} f(\zeta) \frac{I_1 \left[\sqrt{-q(v^2t^2 - (x-\zeta)^2)} \right]}{\sqrt{v^2t^2 - (x-\zeta)^2}} d\zeta .$$

When $q > 0$, (3.87) is the *Klein-Gordon equation* K-GE [3.14], well-known from application to elementary particle and nuclear physics.

For Cauchy initial condition (3.101), the solution of (K-GE) equation can be written as

$$u(x,t) = \frac{f(x - vt) + f(x + vt)}{2} \tag{3.103}$$

$$+ \frac{1}{2v} \int_{x-vt}^{x+vt} g(\zeta) J_0 \left[\sqrt{q(v^2 t^2 - (x - \zeta)^2)} \right] d\zeta$$

$$- \frac{v\sqrt{q}t}{2} \int_{x-vt}^{x+vt} f(\zeta) \frac{J_0 \left[\sqrt{q(v^2 t^2 - (x - \zeta)^2)} \right]}{\sqrt{v^2 t^2 - (x - \zeta)^2}} d\zeta .$$

Both solutions (3.102) and (3.103) exhibit the domains of dependence and influence on *modified telegraph equation* and *Klein-Gordon equation*. These domains, which characterize the maximum speed at which thermal disturbance travel, are determined by the principal parts of the given equation (i.e., the second derivative terms) and do not depend on the lower order terms. It can be concluded that these equations and the wave equation (for $m = 0$) have identical domains of dependence and influence.

3.2.2 Metastable Thermal Quantum States

The concept of metastable states in quantum mechanics dates back to the beginning of the century. The interaction between a quantum system and an external electromagnetic field was described by Planck using the concept of a metastable state. The metastable is distinguished from a stationary state, which is "infinitely" long lived. The metastable state is an unstable state and has finite lifetime [3.27].

In this section, the metastable thermal states created in quantum structures by the ultrashort thermal pulses are investigated. When the ultrashort laser pulses interact with an inhomogeneous quantum structure, the potential energy barriers on the edges of discontinuities influence the thermal energy transport.

The master equation that describes the thermal perturbation propagation is the Klein-Gordon equation (3.87), with $q > 0$, which is the hyperbolic partial differential equation. In complete analogy to the quantum Klein-Gordon equation, the thermal Klein-Gordon equation has the periodic solution – thermal waves. For the thermal wave that propagates in an inhomogeneous structure, the transmission and the reflection phenomena as well as formation of metastable state can be investigated.

The existence of metastable thermal states can be very important in the modeling of the thermal energy dissipation in quantum inhomogeneous structures, in which ultrashort electromagnetic pulses generate the ultrashort thermal harmonic perturbations.

The generalized heat transport equation, which includes the potential, has the form

$$\frac{1}{v^2}\frac{\partial^2 T}{\partial t^2} + \frac{m}{\hbar}\frac{\partial T}{\partial t} + \frac{2Vm}{\hbar^2}T = \nabla^2 T. \tag{3.104}$$

For constant potential barrier $V = V_0$ and in the case of the one-dimensional heat transfer phenomena, (3.104) can be written as

$$\frac{1}{v^2}\frac{\partial^2 T}{\partial t^2} + \frac{m}{\hbar}\frac{\partial T}{\partial t} + \frac{2V_0 m}{\hbar^2}T = \frac{\partial^2 T}{\partial x^2}. \tag{3.105}$$

In the generalized quantum heat transport equation (3.105), we seek a solution in the form

$$T(x,t) = e^{-\frac{t}{2\tau}}u(x.t), \tag{3.106}$$

where the relaxation time, τ, equals

$$\tau = \frac{\hbar}{mv^2}. \tag{3.107}$$

After substituting (3.106) and (3.107) into (3.105), one obtains

$$\frac{1}{v^2}\frac{\partial^2 u}{\partial t^2} - \frac{\partial^2 u}{\partial x^2} + qu(x,t) = 0, \tag{3.108}$$

where

$$q = \frac{2V_0 m}{\hbar^2} - \left(\frac{mv}{2\hbar}\right)^2. \tag{3.109}$$

For $q > 0$, (3.109) is the Klein-Gordon equation. In the following, we will consider the heat transport through the double Dirac delta potential barrier

$$V_0(x) = a\delta(x) + a\delta(x-d), \tag{3.110}$$

where a is the barrier strength and d is the barrier separation. Let $u(x,t)$ determine harmonic plane thermal wave

$$u(x,t) = e^{-i\omega t}\phi(x). \tag{3.111}$$

The function $\phi(x)$ fulfills the time-independent equations

$$\frac{\hbar^2}{2m}\frac{d^2\phi(x)}{dx^2} + \phi(x)\left[\frac{(\hbar\omega)^2}{2mv^2} + \frac{mv^2}{8}\right] = 0 \tag{3.112}$$

when $V_0 = 0$, and

$$\frac{\hbar^2}{2m}\frac{d^2\phi(x)}{dx^2} + \phi(x)\left[\frac{(\hbar\omega)^2}{2mv^2} + \frac{mv^2}{8} - V_0\right] = 0 \tag{3.113}$$

when $V_0 \neq 0$.

In the potential free region, the plane harmonic thermal wave $\phi(x)$ has the form

$$\phi(x) = e^{ikx}. \tag{3.114}$$

After substituting (3.114) into (3.112), one obtains

$$-k^2 + \frac{2m\Omega^2}{\hbar^2} = 0; \qquad k = \left(\frac{2m\Omega^2}{\hbar^2}\right)^{1/2}, \tag{3.115}$$

where

$$\Omega^2 = \frac{mv^2}{8} + \frac{(\hbar\omega)^2}{2mv^2}. \tag{3.116}$$

The solution of (3.113) in three different regions can be expressed as follows:

$$\phi_I(x) = e^{ikx} + Re^{-ikx} \quad x \leq 0,$$

$$\phi_{II}(x) = Ae^{ikx} + Be^{-ikx} \quad 0 \leq x \leq d, \tag{3.117}$$

$$\phi_{III}(x) = Ce^{ikx} \quad x \geq d.$$

Here, R and C are the reflection and transmission amplitudes, respectively. The transmission coefficient equals $|C|^2$. The boundary conditions at $x = 0$ and $x = d$ are used to determine the amplitude C and the transmission coefficient $|C|^2$. At the locations of the delta barriers, the wave functions $\phi(x)$ are continuous, that is

$$\phi_I(0) = \phi_{II}(0), \qquad \phi_{II}(d) = \phi_{III}(d). \tag{3.118}$$

The slope of the wave functions at $x = 0$ and $x = d$ is discontinuous, and it can be shown by carrying out the following integrals:

$$\lim_{\eta \to 0} \int_{-\eta}^{\eta} \left[-\frac{\hbar^2}{2m}\frac{d^2}{dx^2} + a\delta(x) + a\delta(x - d)\right]\phi(x)dx = \lim_{\eta \to 0} \int_{-\eta}^{\eta} \Omega^2\phi(x)dx, \tag{3.119}$$

or

$$-\frac{\hbar^2}{2m}[\phi_{II}'(0) - \phi_I'(0)] + a\phi_I(0) = 0 \tag{3.120}$$

and

$$\lim_{\eta \to 0} \int_{d-\eta}^{d+\eta} \left[-\frac{\hbar^2}{2m}\frac{d^2}{dx^2} + a\delta(x) + a\delta(x - d)\right]\phi(x)dx = \lim \int_{d-\eta}^{d+\eta} \Omega^2\phi(x)dx, \tag{3.121}$$

or

$$-\frac{\hbar^2}{2m}[\phi'_{III}(d) - \phi'_{II}(d)] + a\phi_{III}(d) = 0. \tag{3.122}$$

From (3.120) and (3.122), the discontinuity of the first-order derivatives at $x = 0$, and d can be observed.

By substituting (3.117) into (3.118), (3.120), and (3.122), one obtains

$$1 + R = A + B,$$

$$Ae^{ikd} + Be^{-ikd} = Ce^{ikd}, \tag{3.123}$$

$$-\frac{\hbar^2}{2m}[ikA - Bik - ik + Rik] + a(1 + R) = 0,$$

$$-\frac{\hbar^2}{2m}[ikCe^{ikd} - ikAe^{ikd} + Bike^{-ikd}] + aCe^{ikd} = 0$$

When the dimensionless parameter $\varepsilon = 2ma/k\hbar^2$ is defined, (3.123) can be written as

$$1 + R = A + B,$$

$$Ae^{ikd} + Be^{-ikd} = Ce^{ikd}, \tag{3.124}$$

$$A - B - 1 + R - \frac{\varepsilon}{i}(1 + R) = 0,$$

$$-Ae^{ikd} + Be^{-ikd} + Ce^{ikd}(1 - \frac{\varepsilon}{i}) = 0.$$

After rearrangement of the equations (3.124), transmission amplitude C can be calculated

$$C = \frac{4}{[\varepsilon^2 e^{2ikd} + (i\varepsilon + 2)^2]}. \tag{3.125}$$

With the transmission amplitude C, (formula 3.125), we can define the transmission coefficient, K

$$K = |C|^2. \tag{3.126}$$

For a quantum two-barrier structure (3.110), the residence time r, the time that harmonic thermal pulse spends inside the structure, can be defined as

$$r = \frac{d}{Kv}. \tag{3.127}$$

The residence time, formula (3.127), was first defined by B.R. Mottelson [3.28] in connection with tunneling phenomena in nuclear physics.

In Figs. 3.4 and 3.5, the residence time and the ratio of the residence time r, (3.127), to the relaxation time τ, (3.107), of the thermal harmonic pulse are presented.

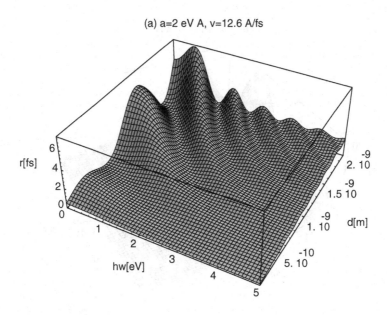

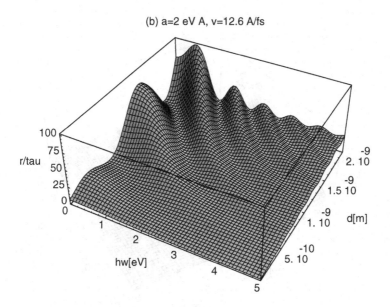

Fig. 3.4. (a) The residence time, r, formula (3.127), for double-barrier structure with $a = 2$ eVA and $v = 1.26$ nm/fs. (b) The ratio r/τ for the double-barrier structure.

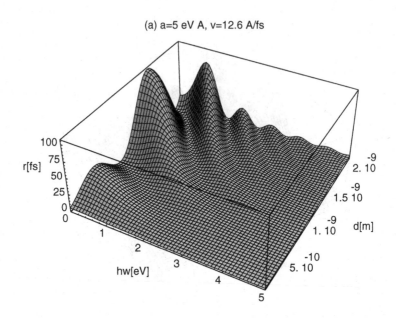

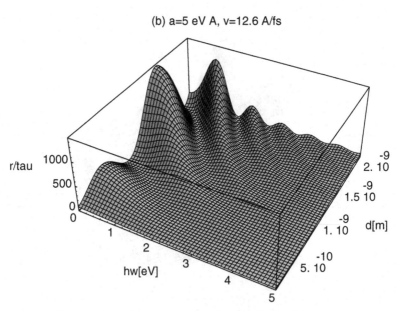

Fig. 3.5. (a) and **(b)** The same as in Figs. 3.4(a) and (b) but for $a = 5$ eVA.

(a) a=2 eV A, v=12.6 A/fs

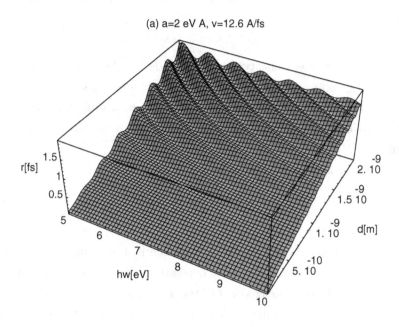

(b) a=2 eV A, v=12.6 A/fs

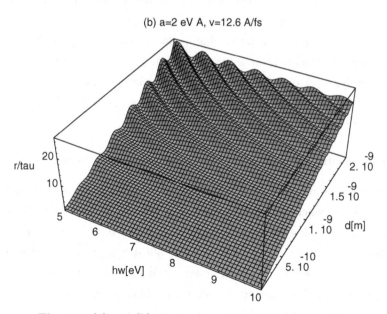

Fig. 3.6. (**a**) and (**b**) The same as in Figs. 3.4(a) and (b).

In Fig. 3.4(a), the residence time was calculated for the quantum structure with potential

$$V(x) = 2\,\text{eVÅ}\delta(x) + 2\,\text{eVÅ}\delta(x - d) \qquad (3.128)$$

and the thermal wave velocity $v = 1.26\,\text{nm/fs} = 12.6\,\text{ÅAfs}$. The resonant peaks corresponding to the metastable states become broadened at higher energy indicating the decreasing lifetime of the metastable thermal states. In Fig. 3.4(b), the ratio r/τ for the same quantum potential (3.128) was calculated.

The results for quantum structure

$$V(x) = 5\,\text{eVÅ}\delta(x) + 5\,\text{eVÅ}\delta(x - d) \qquad (3.129)$$

are presented in Figs. 3.5(a) and (b).

It is quite interesting to observe that for thermal harmonic pulse with an energy $\hbar\omega < 5$ eV, the residence time depends strongly on potential strength parameter a and linear dimension d of the quantum structure. Moreover, for discrete values of thermal harmonic pulse energy, the metastable (with long residence time) states are generated. In these metastable states, the thermal energy is stored inside the structure and is not transmitted outward. In complete analogy to the quantum mechanics for greater energy of the thermal harmonic pulse, the quantum structure is transparent for thermal pulse. The results for $\hbar\omega > 5$ eV are presented in Figs. 3.6(a) and (b). In that case, we observe small "ripples" on the smooth background. The background is described by the traversal time, d/v.

3.2.3 Quantum Heat Transport on the Molecular Scale

Molecular electronics is a new field of science and technology that is evolving from the convergence of ideas from chemistry, physics, biology, electronics, and information technology [3.29, 3.30, 3.31]. It considers, on the one hand molecular materials for electronic/optoelectronic applications, and on the other hand attempts to build electronics with molecules at the molecular level. It is this second viewpoint that concerns us here: describing the heat transport at the level of single or few molecules.

The heat and charge transport phenomena on the molecular scale are the quantum phenomena, and electrons constitute the charge and heat carriers. We argue that to describe the heat transport phenomena on the molecular level, the quantum heat transport equation is a natural reference equation.

The quantum heat transport equation for the atomic scale has the form (3.20, 3.21)

$$\tau\frac{\partial^2 T}{\partial t^2} + \frac{\partial T}{\partial t} = D\nabla^2 T\,, \tag{3.130}$$

where τ is the relaxation time, D is the heat diffusion coefficient, and T denotes temperature. For the atomic scale heat transport, the relaxation time equals

$$\tau = \frac{\hbar}{mv_h^2}, \qquad v_h = \frac{1}{\sqrt{3}}\alpha c\,, \tag{3.131}$$

where m is the electron mass and v_h is the velocity of the heat perturbation. Moreover, on the atomic level, the temperature field $T(r)$ is quantized by a quantum heat of energy, the *heaton*. The *heaton* energy equals

$$E_h = m_e v_h^2\,. \tag{3.132}$$

Due to formula (3.132), the *heaton* energy is the interaction energy of electromagnetic field with electrons (through the coupling constant α).

At the molecular level, we seek energy of interactions of the electromagnetic field with a molecule. This energy is described by the formula, [3.32]

$$e_h^m = \alpha^2 \frac{m_e}{m_p} m_e c^2\,, \tag{3.133}$$

where m_e, m_p are the masses of electron and proton, respectively.

Considering the general formula for *heaton* energy (3.132), one obtains from formula (3.133) for velocity of the thermal perturbation

$$v_h = \alpha c \left(\frac{m_e}{m_p}\right)^{1/2}\,. \tag{3.134}$$

Comparison of formulas (3.131) and (3.134) shows that v_h scales with ratio $(m_e/m_p)^{1/2}$ when the atomic scale is changed to the molecular scale; v_h is the Fermi velocity for molecular gas.

Quantum heat transport equation (3.130) has as a solution, for short time scale (short in comparison to relaxation time τ), the heat waves that propagate with velocity v_h. One can say that on the molecular level, the heat waves are slower in comparison to the atomic scale.

From formulas (3.131) and (3.134), the relaxation time can be calculated

$$\tau = \frac{m_p}{m_e}\frac{\hbar}{m_e c^2 \alpha^2}\,. \tag{3.135}$$

It occurs that relaxation time on the molecular scale is longer (ratio m_p/m_e) than the atomic relaxation time. For standard values of the constants of Nature

$$\alpha = \frac{1}{137}, \qquad m_e = 0.511 \, \text{MeV}/c^2, \qquad m_p = 938 \, \text{MeV}/c^2, \quad (3.136)$$

one obtains the following numerical values for v_h, τ and E_h: $v_h = 0.05 \, \text{nm/fs}$, $\tau = 44 \, \text{fs}$, and $E_h = 10^{-2} \, \text{eV}$. With those values of v_h and τ, the mean free path

$$\lambda = \tau v_h \qquad (3.137)$$

can be calculated and $\lambda = 2.26 \, \text{nm}$. It is interesting to observe that in the structure of the biological cells, some elements have a dimension of the order the nanometer [3.29].

With the help of the *heaton* energy, one can define the *heaton temperature*, i.e. the characteristic temperature of the heat transport on the molecular scale, viz.:

$$T_m = \alpha^2 \frac{m_e}{m_p} m_e c^2 \, 1.16 \, 10^{10} \, \text{K} \approx 10^{-3} \alpha^2 m_e c^2 \sim 316 \, \text{K} \,. \qquad (3.138)$$

This defines what we generally term "room temperature." At temperatures far below T_m, the hydrogen bond becomes very rigid and the flexibility of atomic configurations is weakened. Most substances are liquid or solid below T_m. Biology occurs in environments with ambient temperature within an order or magnitude or so of T_m.

In Figs. 3.7 and 3.8, the results of the theoretical calculations for the quantum heat transport on the molecular scale are presented. In Fig. 3.7(a), the solution of Eq. (3.130) in one-dimensional case

$$\tau \frac{\partial^2 T}{\partial t^2} + \frac{\partial T}{\partial t} = D \frac{\partial^2 T}{\partial x^2} \,, \qquad (3.139)$$

for the following input parameters $v_h = 0.05 \, \text{nm/fs}, \tau = 44 \, \text{fs}, T_0 = 300 \, \text{K}$ (initial temperature) and Δt duration of laser pulse $= 0.2 \, \tau$ is presented.

In Fig. 3.7(b), the solution of quantum parabolic heat transport equation (QHT) (Fourier equation)

$$\frac{\partial T}{\partial t} = D \frac{\partial^2 T}{\partial x^2} \qquad (3.140)$$

with the same input parameters is presented.

In Figs. 3.8(a) and (b), the solutions of (3.139), (3.140) for the same input parameter but for $\Delta t = \tau$ are presented.

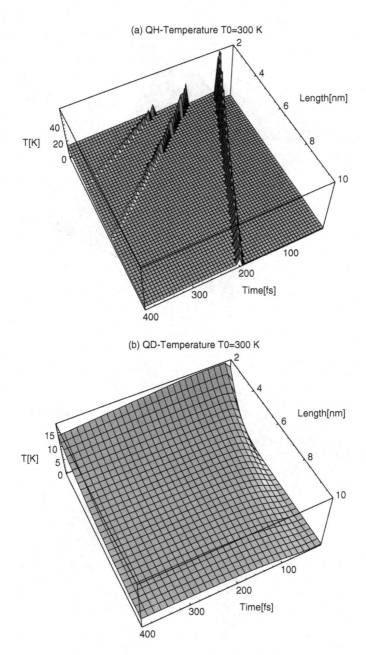

Fig. 3.7. (a) The solution of QHT equation (3.139) for the following input parameters $v_h = 5\,10^{-2}\,\text{nm/fs}, \tau = 44\,\text{fs}, T_0 = 300\,\text{K}$, and $\Delta t = 0.2\tau$. **(b)** The solution of QPT (3.140) with the same input parameters.

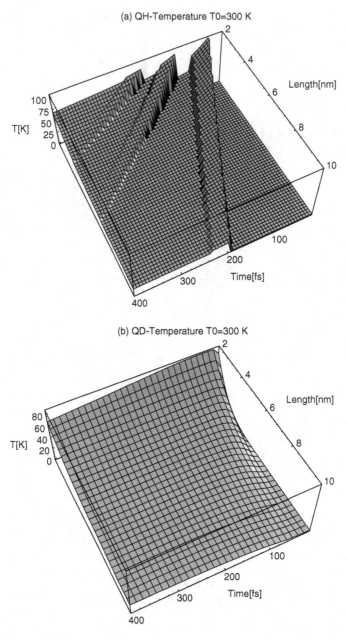

Fig. 3.8. (a) and (b) The same as in Fig. 3.7(a) and (b) but for $\Delta t = \tau$.

From the analysis of the solutions of hyperbolic and parabolic quantum heat transport equation, the following conclusions can be drawn. In the case of QHT equation, the thermal wave dominates the heat transport for $\Delta t = 0.2\ \tau,\ \tau$. The finite value of v_h involves the delay time for the response of the molecular system on the initial temperature change. In the case of QPT, instantaneous diffusion heat transport is observed. From the technological point of view, the strong localization of thermal energy in the front of thermal wave is very important.

References

3.1. C. Catteneo: Atti Sem. Mat. Fis. Univ. Modena **3**, 3 (1948); C. R. Acad. Sc. Paris **247**, 431 (1958)

3.2. A.V. Luikov: *Analytical Heat Diffusion Theory* (Academic Press Inc. New York 1968)

3.3. D.D. Joseph, L. Preziosi: Rev. Mod. Phys. **61**, 41 (1989); **62**, 375 (1990)

3.4. D. Jou, J. Casas-Vázques, G. Lebon: *Extended Irreversible Thermodynamics* (Springer-Verlag Berlin 1993)

3.5. M. Kozlowski: Nucl. Phys. **A492**, 285 (1989)

3.6. E. Nelson: Phys. Rev. **150**, 1079 (1966)

3.7. D.H. Perkins: *Introduction to High Energy Physics* (Addison-Wesley Menlo Park California 1987)

3.8. M. Ivanov, P.B. Corkum, T. Zuo, A. Bandrauk: Phys. Rev. Lett **74**, 2933 (1995)

3.9. J. Marciak-Kozlowska, M. Kozlowski: Lasers Eng. **6**, 141 (1997)

3.10. B.B. Mandelbrot: *The Fractal Geometry of Nature* (W.H. Freeman and Company San Francisco 1989)

3.11. L. F. Abbot, M. B. Wise: Am. J. Phys. **19**, 37 (1981)

3.12. B. Gaveau, T. Jacobson, M. Kac, L.S. Schulman: Phys. Rev. Lett. **53**, 419 (1984)

3.13. S.C. Tiwari: Phys. Lett. A **133**, 279 (1988)

3.14. S.C. Tiwari: Phys. Essays **2**, 31 (1989)

3.15. T. Jacobson, L.S. Schulman: J. Phys. A Gen. **17**, 375 (1984)

3.16. E. Zauderer: *Partial Differential Equations of Applied Mathematics* (J. Wiley and Sons, New York, 1989).

3.17. P. Debye, E. Hückl: Phys. Z. **24**, 185 (1923)

3.18. R. Balian: *From Microphysics to Macrophysics*, vol. **II** (Springer-Verlag Berlin 1992)

3.19. B. Povh, K. Rith, Ch. Scholz, F. Zetsche: *Particles and Nuclei* (Springer-Verlag Berlin 1995)

3.20. T.H.R. Skyrme: Proc. R. Soc. London, Ser. A **247**, 260 (1958) Ch. Nayak, F. Wilczek: Phys. Rev. Lett. **77**, 44 (1996)

3.21. M.D. Perry, G. Mourou: Science **264**, 917 (1994)

3.22. T. Ditmire, T. Donnelly, R.W. Falcone, M.D. Perry: Phys. Rev. Lett. **75**, 3122 (1995)

3.23. Y.L. Shao: Phys. Rev. Lett. **77**, 3343 (1995)

3.24. T. Ditmire: Nature **386**, 54 (1997)

3.25. J. Marciak-Kozlowska, M. Kozlowski: Lasers Eng. **5**, 79 (1996)

3.26. H.S. Carslaw, J.C. Jaeger: *Operational Methods in Applied Mathematics* (Oxford University Press Oxford 1953)

3.27. C. Cohen-Tannoudji, B. Diu, F. Laloe: *Quantum Mechanics* (J. Wiley and Sons New York 1977)

3.28. B.R. Mottelson: The Study of the Nucleus as a Theme in Contemporary Physics. In *The Lesson of Quantum Theory*, ed by J. de Boer, E. Dal, O. Ulfbeck (North Holland Physics Publishing Amsterdam 1986)

3.29. J.R. Barker, *Molecular Electronics – Science and Technology*, ed by A. Aviram (Engineering Foundation 1989) p 213

3.30. J.R. Barker: *Parallel Processing in Neural Systems and Computer*, ed by R. Eckmiller, G. Hartmann and G. Hauske (North-Holland Physics Publishing Amsterdam 1990) p. 519.

3.31. J.C. Diels, W. Rudolph: *Ultrashort Laser Pulse Phenomena* (Academic Press Inc. San Diego 1996)

3.32. Fang Li Zhi, Li Shu Xian: *Creation of the Universe* (World Scientific Singapore 1989)

3.33. M. Kozlowski, J. Marciak-Kozlowska: Hadronic Journal **20**, 289 (1997)

3.34. NA44 Collaboration: Phys. Rev. Lett. **78**, 2080 (1997)

4

Application of the Quantum Heat Transport Equation

4.1 The Pauli-Heisenberg Model

4.1.1 Electron Thermal Relaxation in Metallic Nanoparticles

Clusters and aggregates of atoms in the nanometer size (currently called nanoparticles) are systems intermediate in several respects, between simple molecules and bulk materials, and have been objects of intensive work [4.1]–[4.3]. The main motivation for the growing interest in these systems is related to the possibility of tailoring, to a considerable extent, their physical behavior on the basis of the size [4.4, 4.5].

In paper [4.1], the novel experimental approach to investigate the different mechanism leading to the electron thermalization in nanoparticles was presented. The femtosecond pump-probe measurements on gallium nanoparticles in both the liquid and solid phases were performed. The samples were prepared by evaporation-condensation of high-purity gallium in ultrahigh vacuum on sapphire substrates. The nanoparticle shape was of a truncated sphere. The measurements were performed on three gallium samples with radii r = 5 nm, 7 nm and 9 nm. Transient transmissivity and reflectivity measurements were performed by using a standard pump-probe configuration. The laser system consists of a Ti: sapphire laser with chirped pulse amplification that provides pulses of 150 fs duration at 780 nm with an energy up to 750 μJ at 1 kHz repetition rate.

The main results of the paper [4.1] are: (i) the temporal behavior of the electron energy relaxation is similar in both phases, (ii) the time constant for thermal relaxation is of the order 600–1600 fs.

In this section, we investigate the thermal relaxation phenomena in the nanoparticles in the frame of quantum heat transport equation formulated in papers [4.6, 4.7]. In paper [4.6], the thermal inertia of materials heated with laser pulses faster than the characteristic relaxation time was investigated. It was shown that in the case of the ultrashort laser pulses, the hyperbolic heat conduction (HHC) must be used. For Ga nanoparticles, the mean free path of the electron is larger than the maximum radius of the nanoparticles [4.1]. Moreover, the mean free path is of the order, the de Broglie wave length. In that case, the classical hyperbolic heat conduction equation (HHT) must be replaced by quantum hyperbolic heat transport equation (QHT) (3.20)

$$\tau^e \frac{\partial^2 T^e}{\partial t^2} + \frac{\partial T^e}{\partial t} = \frac{\hbar}{m_e} \nabla^2 T^e, \tag{4.1}$$

where T^e denotes the temperature of the electron gas in nanoparticle, τ^e is the relaxation time, and m_e denotes the electron mass. The relaxation time τ^e is defined as

$$\tau^e = \frac{\hbar}{m_e v_h^2}, \tag{4.2}$$

where v_h is the thermal pulse propagation velocity

$$v_h = \frac{1}{\sqrt{3}} \alpha c. \tag{4.3}$$

In formula (4.3), α is the fine-structure constant, $\alpha = e^2 / \hbar c$, and c denotes the light velocity in vacuum. Both parameters τ^e and v_h completely characterize the thermal energy transport on the atomic scale and can be named as "*atomic relaxation time*" and "*atomic*" *heat velocity*.

Both τ^e and v_h are built up from constant of Nature, $e, \hbar, m_e, c$. Moreover, on the atomic scale there is no shorter time period than τ_e and smaller velocity built from constants of Nature. In consequence, one can name τ^e and v_h as *elementary relaxation time* and *elementary velocity*, which characterize heat transport in the elementary building block of matter, the atom.

In the following, starting with elementary τ^e and v_h, we intend to describe thermal relaxation processes in nanoparticles that consist of N atoms (molecules) each with elementary τ^e and v_h. To that aim, we use the Pauli-Heisenberg inequality [4.8]

$$\Delta r \Delta p \geq N^{\frac{1}{3}} \hbar, \tag{4.4}$$

where r denotes characteristic dimension of the nanoparticle, and p is the momentum of energy carriers. The Pauli-Heisenberg inequality expresses the basic property of the N-fermionic system. In fact, compared with the standard Heisenberg inequality

$$\Delta r \Delta p \geq \hbar, \tag{4.5}$$

we notice that in this case the presence of the large number of identical fermions forces the system either to become spatially more extended for fixed typical momentum dispension or to increase its typical momentum dispension for a fixed typical spatial extension. We could also say that for a fermionic system in its ground state, the average energy per particle increases with the density of the system.

A picturesque way of interpreting the Pauli-Heisenberg inequality is to compare (4.4) with (4.5) and to think of the quantity on the right-hand side of it as the "effective fermionic Planck constant"

$$h^f(N) = N^{\frac{1}{3}}\hbar. \tag{4.6}$$

We could also say that antisymmetrization, which typifies fermionic amplitudes, amplifies those quantum effects that are affected by Heisenberg inequality. It does so to a degree that becomes significant if the number N of identical fermions is large [4.8].

According to formula 4.6, we recalculate the relaxation time τ, formula (4.2) and thermal velocity v_h, formula (4.3) for nanoparticle consisting of N fermions

$$\hbar \to \hbar^f(N) = N^{\frac{1}{3}}\hbar \tag{4.7}$$

and obtain

$$v_h^f = \frac{e^2}{\hbar^f(N)} = \frac{1}{N^{\frac{1}{3}}}v_h, \tag{4.8}$$

$$\tau^f = \frac{\hbar^f}{m(v_h^f)^2} = N\tau. \tag{4.9}$$

Number N particles in nanoparticle (sphere with radius r) can be calculated according to the formula (we assume that density of nanoparticle does not differ too much from bulk material)

$$N = \frac{\frac{4\pi}{3}r^3 \varrho AZ}{\mu} \tag{4.10}$$

and for spherical with semiaxes a, b, c

$$N = \frac{\frac{4\pi}{3}abc\varrho AZ}{\mu}, \tag{4.11}$$

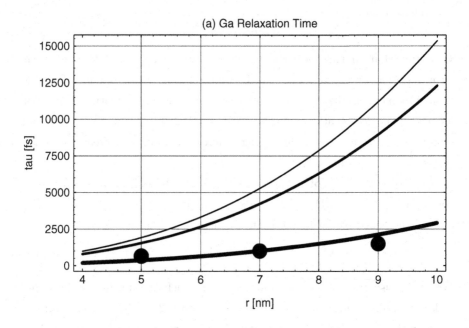

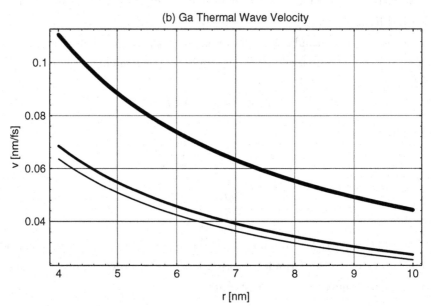

Fig. 4.1. (a) Experimentally determined relaxation times for Ga nanoparticles [4.1]. Curves denote the calculated relaxation times for $d = 0.19$ (—), $d = 0.8$ (—), and $d = 1$ (—). (b) Calculated thermal wave velocities for nanoparticles with $d = 0.19$(—), $d = 0.8$ (—), and $d = 1$ (—).

where ϱ is the density of the nanoparticle, A is the Avogadro number, μ is the molecular mass of particles in grams, and Z is the number of the valence electrons.

With formulas (4.8) and (4.9), we calculate de Broglie wavelength λ_B^f and mean free path λ_{mfp}^f for nanoparticles

$$\lambda_B^f = \frac{\hbar^f}{mv_{th}^f} = N^{\frac{2}{3}}\lambda_B, \tag{4.12}$$

$$\lambda_{mfp}^f = v_{th}^f \, \tau_{th}^f = N^{\frac{2}{3}}\lambda_{mfp}, \tag{4.13}$$

where λ_B and λ_{mfp} denote the de Broglie wave length and mean free path for heat carriers in nanoparticles.

In the following, we will study the thermal relaxation process in gallium particles [4.1]. For Ga, density $\varrho = 5.9$ g/cm^3 and $\mu = 70$ g. In Fig. 4.1, we present the calculation of the relaxation time τ^f formula (4.9) and thermal wave velocity v_{th}^f (4.8) for Ga nanoparticles, when axes $a = b = r$ and $c = dr, d \leq 1$ (symmetric spheroid). In that case

$$N = \frac{\frac{4\pi}{3}dr^3\varrho AZ}{\mu}. \tag{4.14}$$

Figure 4.1(a) shows the calculations of the τ^f for $d = 0.19, 0.8$ (spheroid shape) and for $d = 1$ (sphere). As can be seen, the fairly good agreement is obtained for spheroid with semiaxes $a = b = r$ and $c = 0.19r$. Figure 4.1(b) shows the calculated velocities of the thermal waves. The reduction of the thermal wave velocity is caused by electron-electron scattering and reflecting of the thermal wave from the surface of the nanoparticles.

4.1.2 Velocity Spectra of the Relativistic Electrons

The implementation of chirped pulse amplification (CPA) systems in high power lasers has made available new intensity regimes previously inaccessible in the laboratory. At intensities of 10^{18} W/cm^2, the electron oscillatory velocity for $1\,\mu$m radiation becomes relativistic and the radiation pressure reaches 300 Mbar. Interesting, new physical phenomena have been predicted in this regime, such as emission high energetic electrons, ions, and MeV X-ray.

In the physical picture of a relativistic gas, we think of the *world lines* of the particles as a discrete complex. For a particle with mass m, the four momentum relationships M_r fulfills the relations $M_r M_r = -m^2$ for massive particles and $M_r M_r = 0$ for photons. If η is the number of world lines for particles that cross the space volume element $dS = dx_1 dx_2 Dx_3$ and momentum volume $d\Omega = dM_1 dM_2 dM_3$, then η equals [4.9]

$$\eta = N \, DS \, D\Omega \,, \tag{4.15}$$

N being independent of the sizes of dS and $d\Omega$. We call N the relativistic distribution function. For massive particles with mass m, the four momentum components are related to three velocity component u_S by the equations

$$M_S = \frac{m\gamma u_S}{c}, \qquad \gamma = \left(1 - \frac{u^2}{c^2}\right)^{-1/2}, \tag{4.16}$$

where c is the velocity of light. The relation between $d\Omega$ and $dU = du_1 du_2 du_3$ reads

$$d\Omega = \frac{m^3 \gamma^5}{c^3} \, dU. \tag{4.17}$$

In the rest frame of the relativistic gas container, the distribution function N has the form

$$N = \frac{N_0 T^{-1}}{4\pi m^2 K_2(\frac{m}{T})} \exp\left[-\frac{m\gamma}{T}\right]. \tag{4.18}$$

In formula (4.18), N_0 is the *numerical density* (number of particles per volume in the rest frame of the container), T is the relativistic gas temperature (in energy units), and $K_2(m/T)$ is the modified Bessel function of the second kind. From formulas (4.15) and (4.18), the number of particles in $dSdU$, in the rest frame of gas is

$$d\eta = \frac{N_0 m T^{-1}}{4\pi K_2(\frac{m}{T})} c^{-3} \gamma^5 \exp\left[-m\gamma T^{-1}\right] dSdU. \tag{4.19}$$

The number of particles in the volume dS in the range $(u, u+du)$ is obtained by equating $4\pi u^2 \, dU$ to dU and is therefore

$$d\eta = N_0 \frac{m T^{-1}}{K_2(\frac{m}{T})} c^{-3} \gamma^5 u^2 \exp\left[-m\gamma T^{-1}\right] dSdU. \tag{4.20}$$

If we define $\beta = u/c$, then formula (4.20) reads

$$d\eta = \frac{N_0 m T^{-1}}{K_2(\frac{m}{T})} \gamma^5 \beta^2 \exp\left[-m\gamma T^{-1}\right] dSd\beta, \tag{4.21}$$

and describes the number of particles in the volume dS in the range $(\beta, \beta+d\beta)$.

For low temperatures, i.e. for $mT^{-1} \gg 1$, $K_2(m/T)$ has an approximate form

$$K_2(x) \cong \sqrt{\frac{\pi}{2x}} e^{-x}; \qquad x = mT^{-1}. \tag{4.22}$$

Considering formula (4.22), Eq. (4.21) reads

$$d\eta = \frac{4\pi}{\sqrt{8\pi^3}} N_0 \left(\frac{mc^2}{T}\right)^{3/2} \gamma^5 \beta^2 \exp\left[-\frac{mc^2(\gamma - 1)}{T}\right] d\beta, \tag{4.23}$$

where $E = mc^2(\gamma - 1)$ is the kinetic energy of the particle with mass m. Formula (4.23) is the relativistic analog of the Maxwell formula for nonrelativistic particles, i.e. for particles with $T \ll mc^2$.

Because the exponential falls off rapidly with increasing T, u/c is small for vast majority of the particles, and for them we may replace γ by 1 and E by nonrelativistic kinetic energy $E \sim 1/2mu^2$; then (4.23) reads

$$d\eta = \frac{4\pi}{\sqrt{8\pi^3}} N_0 \left(\frac{mc^2}{T}\right)^{3/2} \beta^2 \exp\left[-\frac{m\beta^2}{2T}\right] d\beta. \tag{4.24}$$

This is precisely a Maxwellian distribution and so we are assured that T is, in fact, the absolute temperature in the ordinary sense.

Note, however, that we obtained (4.24) from (4.23) merely in order to make contact with Maxwellian theory; formula (4.23) is a much better expression relativistically, because it corresponds to the range $(0, c)$ for u, whereas (4.24) corresponds to the range $(0, \infty)$. On account of this difference in ranges, it is not quite correct to say that for low temperatures the relativistic theory is reduced to the Maxwellian case. In the case of high temperatures $T \sim mc^2$, we derive from formula (4.21)

$$d\eta = \frac{N_0}{2} \left(\frac{mc^2}{T}\right)^3 \gamma^5 \beta^2 \exp\left[-\frac{m\gamma c^2}{T}\right] dS d\beta, \tag{4.25}$$

for $K_2(x) \sim 2x^{-2}$ when x is small.

Recent advances in high intensity sub-picoseconds laser technology enables new regimes of hot dense matter to be investigated [4.10]–[4.15]. Dense high temperature plasmas are typically studied by X-ray spectroscopy. Time of flight (TOF) measurements have been used to determine suprathermal electron temperatures of plasmas produced by lasers with pulse length 1 ps [4.16] to 1 ns [4.17]. Spatially and temporally averaged X-ray spectra of sub-ps laser produced plasmas have shown electron temperatures of a few hundred eV [4.18]. Spatially and temporally localized measurements of 500 eV electron temperatures were reported in [4.19].

For electrons with temperatures of the order hundred eV (10^6 K), the quantum heat transport equation (QHT)

$$\tau \frac{\partial^2 T}{\partial t^2} + \frac{\partial T}{\partial t} = \frac{\hbar}{m} \nabla^2 T \qquad (4.26)$$

must be used for the description of the transport phenomena.

The solution of the QHT shows the temperature oscillations for the time period of the few relaxation times [3.25]. The predicted oscillations propagate as a thermal wave with velocity $v_{th} \sim \alpha c$ where α is the fine structure constant. The temperature oscillations as well as the thermal waves are a true relativistic effect. For a nonrelativistic transport description, c is infinite and relaxation time $\tau = 0$. In that case, the quantum heat transport is the parabolic equation (QPT), viz.

$$\frac{\partial T}{\partial t} = \frac{\hbar}{m} \frac{\partial^2 T}{\partial t^2}, \qquad (4.27)$$

and oscillations as well as the thermal waves are completely attenuated. The relativistic oscillations of the temperature strongly influence the velocity of the emitted particles.

Subsequently, the electron spectra emitted after irradiation of 5 nm Au film will be calculated. The following input parameters are assumed: initial energy density 10^{16} W/cm^2 and 15% of the light absorbed. The thermal wave velocity $v_h = 0.15\,\mu$m/ps [3.12] and relaxation time $\tau = 20$ fs.

Figure 4.2(a) shows the solution of QHT equation (4.26) and Fig. 4.2(b) the solution of QPT. In Fig. 4.3, the velocity spectra of the particles emitted when the temperature is calculated from QHT equation for $t = 3.3\,10^{-2}$ ps, 10^{-1} ps, and $1.7\,10^{-1}$ ps is showed. Figure 4.3(a) shows the velocity spectra calculated from formula (4.24), Maxwell-Boltzmann distribution function, and Fig. 4.3(b) the relativistic distribution function, formula (4.21). The Maxwellian distribution function gives the wrong result for particle spectra when the temperature T lies in the range of electron mass ($m_e = 0.511$ MeV) as the one obtains $u/c > 1$ for emitted electrons. On the other hand, the relativistic distribution function (4.21) shows the localization of the electron velocity in the vicinity $u \rightarrow c$. Moreover, the velocity spectra oscillate as the function of time.

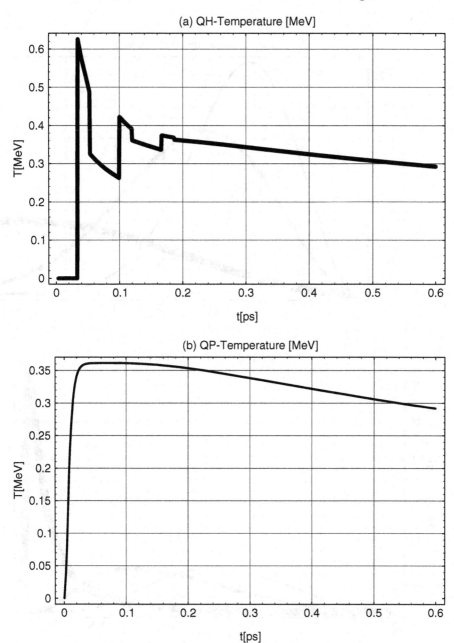

Fig. 4.2. (a) The solution of the QHT equation (4.26) "QH-Temperature" for $v_h = 1.5 \, 10^2$ nm/ps, relaxation time 20 fs, Δt = pulse duration = τ, energy density 10^{16} W/cm^2. (b) The solution of QPT equation (4.27) "QP-Temperature" for the same value of v_h, τ, and energy density.

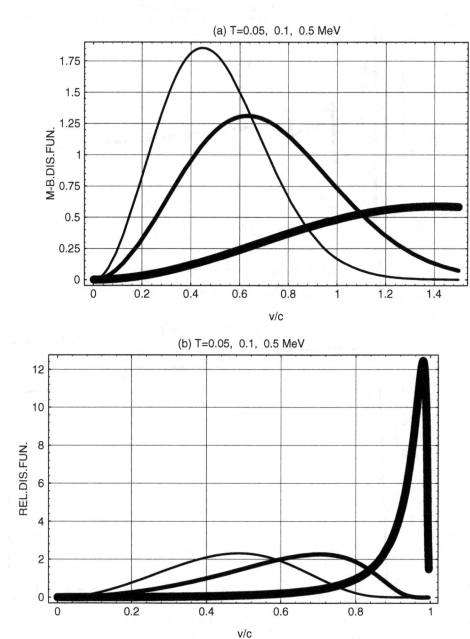

Fig. 4.3. (a) The Maxwellian distribution function (4.24) for $t = 3.3\,10^{-2}\,\text{ps}$ (—), $t = 10^{-1}\,\text{ps}$ (—), and $1.7\,10^{-1}\,\text{ps}$ (—). (b) The relativistic distribution function (4.21) for the same values of time.

4.2 Ballistic and Diffusion Heat Transport

4.2.1 Ballistic and Diffusion Thermal Pulse Propagation in the Attosecond Time Domain

For the first few decades after the invention of the laser in 1960, the record for the shortest laser pulses fell by a factor of two every three years or so. Each development provided new insight into the microworld of atoms, molecules, and solids. In 1986, however, this trend essentially stopped when the pulse length reached 6 femtoseconds. For visible light this corresponds to just three oscillations of the electromagnetic field in the laser.

Since the mid-1980s, there have been many advances in laser science, but the minimum pulse duration has decreased only slightly. In order to break significantly the current record, a 4.5 femtosecond pulse from the laser with a wavelength of 800 nanometers, a completely different approach is needed. Physicists at the Foundation for Research and Technology – Hellas (FORTH) on Crete have recently demonstrated one such approach [4.20].

One place to look for inspiration is the method currently used to measure femtosecond pulses. Basically, we ask the pulse to measure itself. Practically, this means that we take the pulse, split it in two, delay the replicas in the arms of an interferometer by amounts that we can control, and then direct each replica onto a material with nonlinear optical properties. If the two arms have exactly the same length, the pulses arrive at the same time and their intensity is higher than if the path is unbalanced and one pulse arrives before the other. Observing the nonlinear response allows us to measure the extent of the overlap as we change the difference between the two path lengths, Δx. The pulse duration τ, is simply $\tau = \Delta x/c$.

The FORTH group produced replica pulses by splitting the pulse when it is easy to do so – before the high harmonics are produced. And they use the ionizing gas, which produces the high harmonics for a second purpose. It also serves as the nonlinear medium needed to measure the length of the pulses in the train. In a result, an isolate less than 100 as (1 as = 1 attosecond = 10^{-18} s) sharp feature, indicative for the production of the trains of sub-fs XUV pulses, is clearly observable in the resulting temporal trace [4.20].

In the paper [4.20], the time structure with the line width of the order 100 attoseconds was obtained. This debut of attosecond science opens new avenues for investigating atomic and molecular structures. As was shown in papers [4.21] and [4.22] for these two levels of complexity the relaxation times are of the order $\tau_a \sim 70$ attosecond [4.21] and $\tau_m \sim 10^3$ attosecond [4.22]. In both cases, the line width is of the order of/or longer than the relaxation times. For these circumstances, the parabolic Fourier equation cannot be used [4.21, 4.22]. Instead, the new equation, QHT is the valid equation. The QHT can be written as

$$\frac{1}{v^2}\frac{\partial^2 T}{\partial t^2} + \frac{m}{\hbar}\frac{\partial T}{\partial t} = \frac{\partial^2 T}{\partial x^2}. \tag{4.28}$$

In equation (4.28), v denotes the velocity of the thermal pulse propagation, and m is the heat carrier mass. In this paragraph we will consider the Fermi gas of electrons and $m = 0.511$ MeV, $v = \frac{1}{\sqrt{3}}\alpha c$, where α is the electromagnetic fine-structure constant, and $c =$ light velocity. The Cauchy initial condition for (4.28) can be written as

$$T(x,0) = 0, \qquad T(0,t) = f(t). \tag{4.29}$$

For initial conditions (4.29) the solution of (4.28) has the form [3.26]

$$T(x,t) = \left\{ f\left(t - \frac{x}{v}\right)e^{-\frac{\varrho x}{v}} \right. \tag{4.30}$$

$$\left. + \frac{\sigma x}{v}\int_{\frac{x}{v}}^{t} f(t-y)e^{-y\varrho}\frac{I_1\left[\sigma\left(y^2 - \frac{x^2}{v^2}\right)^{\frac{1}{2}}\right]}{\left(y^2 - \frac{x^2}{v^2}\right)^{\frac{1}{2}}}\,dy \right\}H\left(t - \frac{x}{v}\right).$$

In formula (4.30), $\varrho = \sigma = \frac{1}{2\tau}$, $\tau = \frac{\hbar}{mv^2}$ and $H(t - \frac{x}{v})$ is the UnitStep function:

$$H\left(t - \frac{x}{v}\right) = 1 \qquad \text{for} \qquad t \geq \frac{x}{v},$$

$$H\left(t - \frac{x}{v}\right) = 0 \qquad \text{for} \qquad t < \frac{x}{v}. \tag{4.31}$$

As can be seen from formula (4.30), the temperature field $T(x,t)$ has two components, $T_B(x,t)$ – ballistic and $T_D(x,t)$ – diffusion, i.e.,

$$T_B = f\left(t - \frac{x}{v}\right)e^{-\frac{\varrho x}{v}}H\left(t - \frac{x}{v}\right), \tag{4.32}$$

$$T_D = \left\{ \frac{\sigma x}{v}\int_{\frac{x}{v}}^{t} f(t-y)e^{-y\varrho}\frac{I_1\left[\sigma\left(y^2 - \frac{x^2}{v^2}\right)^{\frac{1}{2}}\right]}{\left(y^2 - \frac{x^2}{v^2}\right)^{\frac{1}{2}}}\,dy \right\}H\left(t - \frac{x}{v}\right).$$

In the Figs. 4.4–4.9, the solutions of Eq. (4.30) for the initial condition (4.29) with [3.31]

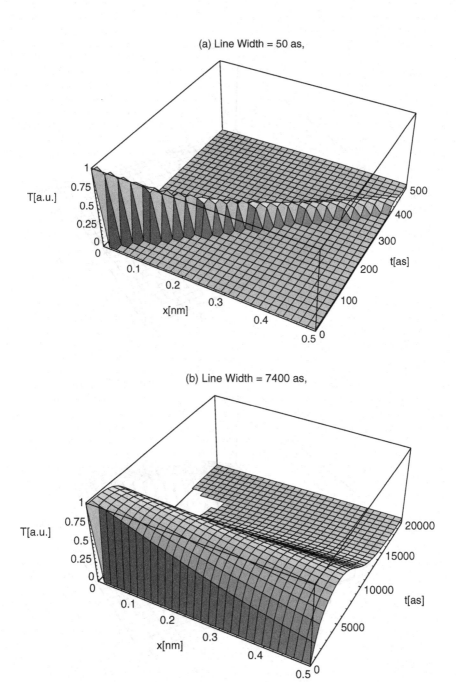

Fig. 4.4. (a) The solution of QHT (4.30) for $t_s = 50$ as. (b) The solution of QHT for $t_s = 7400$ as.

(a) Line Width = 50 as, Ballistic Component

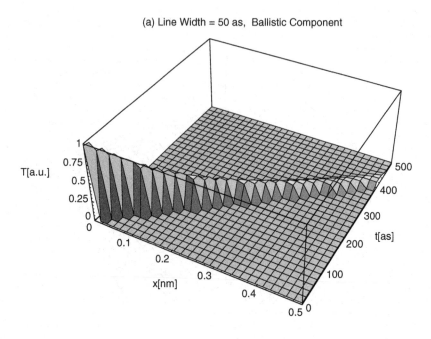

(b) Line Width = 50 as, Diffusion Component

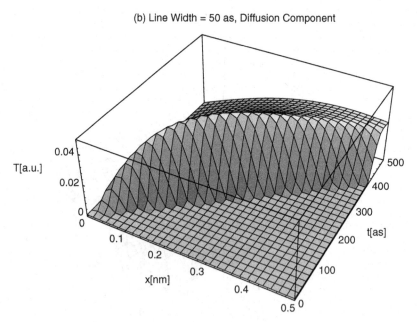

Fig. 4.5. (a) The ballistic component of the heat pulse for $t_s = 50$ as. (b) The diffusion component for $t_s = 50$ as.

(a) Line Width = 74 as, Ballistic Component

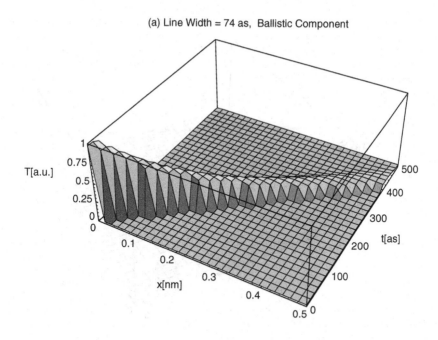

(b) Line Width = 74 as, Diffusion Component

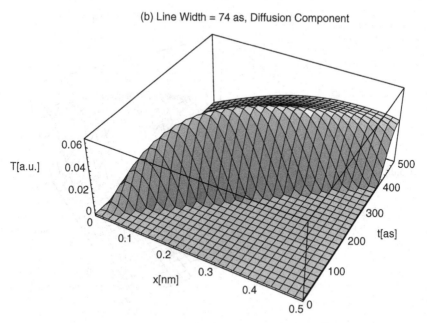

Fig. 4.6. (a) and (b) The same as in Fig. 4.5 but for $t_s = 74$ as.

(a) Line Width = 100 as, Ballistic Component

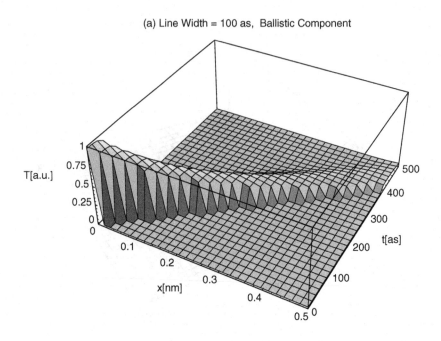

(b) Line Width = 100 as, Diffusion Component

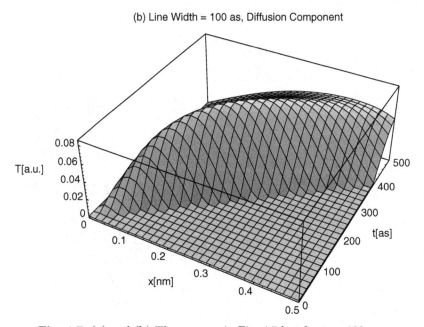

Fig. 4.7. (a) and (b) The same as in Fig. 4.5 but for $t_s = 100$ as.

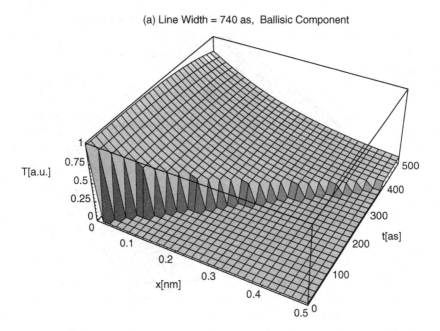

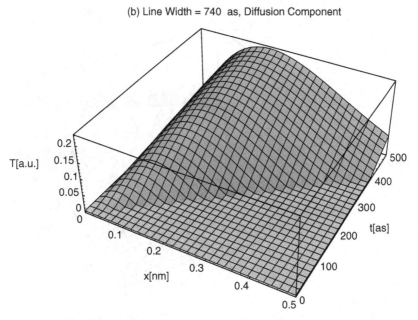

Fig. 4.8. (a) and (b) The same as in Fig. 4.5 but for $t_s = 740$ as.

(a) Line Width = 7400 as, Ballistic Component

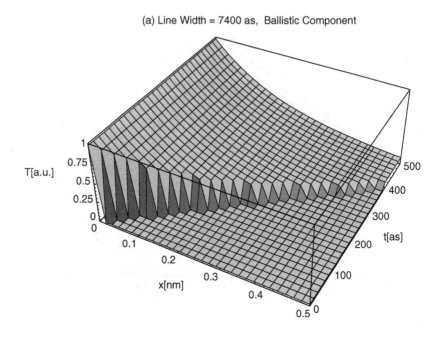

(b) Line Width = 7400 as, Diffusion Component

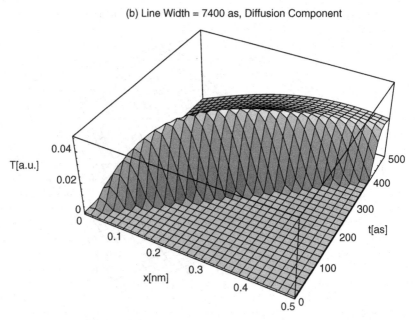

Fig. 4.9. (a) and **(b)** The same as in Fig. 4.5 but for $t_s = 7400$ as.

$$f(t) = Sech^2 \left[\frac{t}{t_s} \right] \tag{4.33}$$

are presented. The numerical integration in the formula (4.30) was performed with the *Mathematica* code, for $t_s = 50$, 74, 100, 740, and 7400 as. In Fig. 4.4, the solution of (4.30) for $t_s = 50$ as (Fig. 4.4(a)) and $t_s = 7400$ as (Fig. 4.4(b)) is presented. The line width $t_s = 50$ as is smaller than the relaxation time $\tau = 74$ as and $t_s = 7400 = 100\tau_s$. The change of the structure of the solutions is evidently seen. Figures 4.5–4.9 represent the analysis of the heat pulse according to formula (4.32). The ballistic (Figs. 4.5(a) to 4.9(a)) and diffusion components (Figs. 4.5(b) to 4.9(b)) have quite different shapes.

The results presented in Figs. 4.4–4.9 describe the heat transport on the atomic scale ($x \sim 0.5$ nm). For times of the order the atomic relaxation time (ballistic propagation), the heat pulse preserves its shape and only the amplitude is diminished due to scattering. On the other hand, for the longer time periods, a new structure develops. Multiple scatterings distort the shapes of the initial pulse. One can say that for $t_s \gg \tau$, the information contained in the initial pulse is lost as the time $\to \infty$.

4.2.2 The Polarization of the Electrons Emitted After Ultrashort Laser Pulse Interaction with Spin Active Solids

Let us consider the quantum heat transport equation for dissipative medium with potential V (3.84)

$$\frac{1}{v^2} \frac{\partial^2 T}{\partial t^2} + \frac{m}{\hbar} \frac{\partial T}{\partial t} + \frac{2Vm}{\hbar^2} T = \frac{\partial^2 T}{\partial x^2} . \tag{4.34}$$

In equation (4.34), v denotes the velocity of heat propagation, m is the mass of heat carrier, and T denotes the temperature. For quantum heat transport equation (4.34), we seek solution in the form

$$T(x,t) = e^{-\frac{t}{2\tau}} u(x,t) , \tag{4.35}$$

where τ denotes the characteristic relaxation time

$$\tau = \frac{\hbar}{mv^2} . \tag{4.36}$$

After substitution of (4.35) into (4.34), one obtains

$$\frac{\partial^2 u(x,t)}{\partial t^2} = v^2 \frac{\partial^2 u}{\partial x^2} + b^2 u(x,t) , \tag{4.37}$$

where

$$b = \sqrt{\left(\frac{mv^2}{2\hbar}\right)^2 - \frac{2Vmv^2}{\hbar^2}}. \tag{4.38}$$

For particles with spin (e.g., electrons), potential V will contain the term describing the spin-orbit interaction. In that case, potential V equals

$$V(r) = V_{\text{central}}(r) + V_{ls}(r)\frac{<ls>}{\hbar^2}, \tag{4.39}$$

where $V_{\text{central}}(r)$ denotes the potential part, which does not depend on spin. The combination of the orbital angular momentum l and electron spin s leads to a total angular momentum $j\hbar = l\hbar \pm \hbar/2$ and hence to the expectation values:

$$<ls> = \begin{cases} \frac{l}{2} & \text{for} \quad j = l + \frac{1}{2}, \\ -\frac{(l+1)}{2} & \text{for} \quad j = l - \frac{1}{2}. \end{cases} \tag{4.40}$$

Substituting formula (4.40) to formula (4.39), one obtains the splitting for potential V:

$$V^+(r) = V_{\text{central}}(r) + V_{ls}(r)\frac{l}{2} \qquad \text{if} \qquad j = l + \frac{l}{2},$$

$$V^-(r) = V_{\text{central}}(r) - V_{ls}(r)\frac{(l+1)}{2} \qquad \text{if} \qquad j = l - \frac{1}{2} \tag{4.41}$$

and for parameter b (formula(4.38))

$$b^+ = \sqrt{\left(\frac{mv^2}{2\hbar}\right)^2 - \frac{2V^+mv^2}{\hbar^2}},$$

$$b^- = \sqrt{\left(\frac{mv^2}{2\hbar}\right)^2 - \frac{2V^-mv^2}{\hbar^2}}. \tag{4.42}$$

One concludes that existence of the spin-orbit term splits the (4.37) into two equations:

$$\frac{\partial^2 u^+(x,t)}{\partial t^2} = v^2\frac{\partial^2 u^+(x,t)}{\partial x^2} + (b^+)^2 u^+(x,t), \tag{4.43}$$

$$\frac{\partial^2 u^-(x,t)}{\partial t^2} = v^2\frac{\partial^2 u^-(x,t)}{\partial x^2} + (b^-)^2 u^-(x,t). \tag{4.44}$$

In the following, we will consider the constant potentials: V_{central} and V_{ls}. The general solution of (4.43) and (4.44) for Cauchy boundary conditions:

$$u(x,0) = f(x), \qquad \frac{\partial u(x,t)}{\partial t}\bigg|_{t=0} = F(x)$$

has the form:

$$u^{+,-}(x,t) = \frac{f(x-vt)+f(x+vt)}{2} + \frac{1}{2}\int_{x-vt}^{x+vt} \Phi^{+,-}(x,t,z)dz,$$

where

$$\Phi^{+,-}(x,t,z) = \frac{1}{v}F(z)J_o\left(\frac{b^{+,-}}{v}\sqrt{(z-x)^2-v^2t^2}\right)$$

$$+ b^{+,-}tf(z)\frac{J_o'\left(\frac{b^{+,-}}{v}\right)\sqrt{(z-x)^2-v^2t^2}}{\sqrt{(z-x)^2-v^2t^2}} \qquad (4.45)$$

and $J_o(z)$ denotes the Bessel function of the first kind. The interaction of laser beam with solid creates hot electrons with temperatures described by formulas (4.35) and (4.45):

$$T^+(x,t) = e^{-\frac{t}{2\tau}}u^+(x,t),$$

$$T^-(x,t) = e^{-\frac{t}{2\tau}}u^-(x,t). \qquad (4.46)$$

The velocity spectra of the emitted hot electrons are described by formula

$$d\eta^{+,-} = N_0\frac{m(T^{+,-}(x,t))^{-1}}{K_2\left(\frac{m}{T^{+,-}(x,t)}\right)}\gamma^5\beta^2\exp\left[-m\gamma(T^{+,-}(x,t))^{-1}\right]dVd\beta. \qquad (4.47)$$

In formula (4.47), $\beta = v/c, \gamma = 1/\sqrt{1-\beta^2}, N_0$ is the numerical density i.e. initial number of particles per volume in the rest frame of the solid, and $K_2[m/T]$ is the modified Bessel function of the second kind. Formula (4.47) describes the number of particles with temperatures T^+, T^- in the volume dV in the range $(\beta, \beta+d\beta)$.

The particles with temperature T^+ have the total angular momentum $j = l + \frac{1}{2}$, and particles with temperature T^- have the total angular momentum $j = l - 1/2$. One can define the degree of the polarization of the emitted electrons (after solid irradiation by an energetic laser beam)

$$P(x,t) = \frac{N^+(x,t)-N^-(x,t)}{N^+(x,t)+N^-(x,t)}, \qquad (4.48)$$

where

$$N^{+,-} = \frac{d\eta^{+,-}}{dV\,d\beta}.$$

Considering formulas (4.47) and (4.48), one concludes that existence of the spin-orbit term in potential V (formula (4.39)) creates the polarization of the emitted electrons. For $V_{ls}(r) = 0$, $N^+(x,t) = N^-(x,t)$ and P(x, t)=0.

4.2.3 Laser Light–Induced π-mesons Emission

In the Rutherford Appleton Laboratory is currently under construction a new PW (petawatt) laser system that will be capable of focusing laser intensities on target of 10^{21} watts per square centimeter. This machine was completed in 2003. The first experiment will be devoted to the pion production (gamma, pion reaction) [4.26]. When high energy *photons* > 1.022 MeV pass through solids, considerable energy is lost by pair creation. Shkolnikov et al. [4.27] have recently calculated that when a sub-terawatt laser interacts with Ta, some $\sim 10^6$ positrons are created per pulse with a positron spectrum that has a maximum energy of about ~ 0.5 MeV. Recently, Karsch et al. [4.26] have estimated that with laser interaction as high as 10^{21} W/cm^2, some 10^9 fast β^- can be generated per pulse. As well as calculating the positron intensity generated by laser of intensity 10^{21} W/cm^2, Karsch et al. [4.26] have also calculated the yields of pions – one of the lightest of the elementary particles. Pion production is energetically possible when γ-rays > 140 MeV interact with solid targets. As was shown in paper [4.26], the yield of relativistic particles, i.e. particles with $E_{kin} > mc^2$, for laser created electrons and pions is extended to 500 MeV and 300 MeV for electrons and pions, respectively. For relativistic particles, the mass depends on velocity of the particle [4.28]

$$m = m_0\gamma, \qquad m_0 = \text{rest mass}, \qquad \gamma = \frac{1}{\sqrt{1 - \frac{v^2}{c^2}}}.$$

However in our earlier model for heat transport in laser-induced reaction [4.7] we assumed $m = m_0$. In this section, we will formulate the extended quantum heat transport equation (EQHT) for relativistic particles with $m = m_0\gamma$. Within frame of the EQHT, the *heaton* energies [4.28] and the spectra of the emitted pions will be calculated.

In paper [4.7] the quantum heat transport equation was derived:

$$\frac{1}{v^2}\frac{\partial^2 T}{\partial t^2} + \frac{m}{\hbar}\frac{\partial T}{\partial t} = \nabla^2 T, \tag{4.49}$$

where T denotes temperature of the heated gas (electrons, nucleons, or quarks), v is the velocity of the thermal disturbance propagation, m is the mass of heat carriers, and $\hbar$ is the Planck constant. When the kinetic energy of the particle $E_k > mc^2$, the mass of the particle depends on the velocity [4.28]:

$$m = m_0\gamma, \tag{4.50}$$

where $\gamma = (1 - v/c^2)^{-1/2}$ and the kinetic energy is defined as

$$E_{\text{kin}} = m_0 c^2 (\gamma - 1) \,. \tag{4.51}$$

Accordingly, the momentum p, $p = m_0 v \gamma$. In formulae (4.50, 4.51), $m_0 = 0.5$ MeV/c^2, 139 MeV/c^2, and 981 MeV/c^2 for electron, *meson* π, and nucleon respectively. Substituting formula (4.50) into formula (4.49), one obtains:

$$\frac{1}{v^2} \frac{\partial^2 T}{\partial t^2} + \frac{m_0 \gamma}{\hbar} \frac{\partial T}{\partial t} = \nabla^2 T \,. \tag{4.52}$$

Table 4.1. Relaxation Times

	Relaxation Time	
	Relativistic	Nonrelativistic
Electromagnetic interaction $\alpha = 1/135$	$\dfrac{\hbar\sqrt{1-\alpha^2}}{m_0 \alpha^2 c^2}$	$\dfrac{\hbar}{m_0 \alpha^2 c^2}$
Strong nucleon-nucleon $\alpha = 0.15$	$\dfrac{\hbar\sqrt{1-\alpha^2}}{m_0 \alpha^2 c^2}$	$\dfrac{\hbar}{m_0 \alpha^2 c^2}$
Strong quark-quark $\alpha \to 1$	$\dfrac{\hbar\sqrt{1-\alpha^2}}{m_0 \alpha^2 c^2}$	$\dfrac{\hbar}{m_0 \alpha^2 c^2}$

In paper [4.7], the *heaton*, quantum of thermal energy, was defined

$$E_h = m v^2 \,. \tag{4.53}$$

Considering formula (4.50), one obtains

$$E_h = m_0 \gamma v^2 \,.$$

To find the physical meaning of the relativistic definition of the *heatons*, we calculate the energy-momentum interval:

$$\frac{E_h^2}{v^2} - p_h^2 = 0 \,. \tag{4.54}$$

It means that for *heatons* we have

$$E_h = p_h v \,, \qquad E_h = p_h \alpha c \tag{4.55}$$

in complete analogy as for *photons*

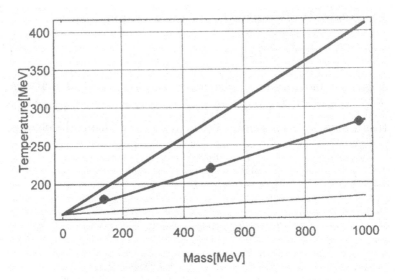

Fig. 4.10. The heaton temperature as the function of the heaton mass. (•) – denotes the masses of π-*meson* (139 MeV/c^2), K-*meson* (490 MeV/c^2), and *proton* (980 MeV/c^2). The line (—) denotes the theoretical calculation (formula (4.64)) for $\alpha =$ 0.15 and (—) for $\alpha = 0.35$, (—) for $\alpha = 0.5$

$$E_{ph} = p_{ph}c. \tag{4.56}$$

As can be easily seen from formula (4.55), the *heaton* velocity is $v = \alpha c$, where α is the strength of the interactions (e.g. electromagnetic $\alpha = 1/137$ or strong $\alpha = 0.15$). Considering the value of v, the equation (4.52) has the form:

$$\frac{1}{\alpha^2 c^2}\frac{\partial^2 T}{\partial t^2} + \frac{m_0}{\hbar\sqrt{1-\alpha^2}}\frac{\partial T}{\partial t} = \nabla^2 T \tag{4.57}$$

and *heaton* energy

$$E_h = \frac{m_0\alpha^2 c^2}{\sqrt{1-\alpha^2}}. \tag{4.58}$$

Formula (4.58) describes the quantum of temperature field $T(\mathbf{r}, t)$ and can be named *heaton* temperature

$$E_h \rightarrow T_h = m_0\alpha^2 c^2 (1 - \alpha^2)^{-1/2}. \tag{4.59}$$

The new result presented by formula (4.59) is that the *heaton* temperature is proportional to *heaton* mass m_0.

Recently, the production of *mesons* (π, K) and nucleons (p) were measured in relativistic heavy ion reaction [4.29]. The temperature of the gases of nucleons and *mesons* were measured. In Fig. 4.10, we present the comparison

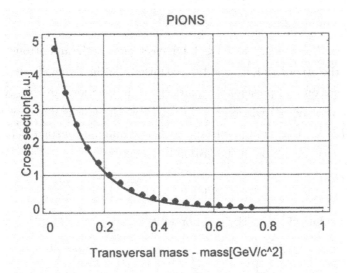

Fig. 4.11. Comparison of the theoretical cross section (4.61) and experimental data (•) [4.29].

of the temperatures for (π, K, p) and theoretical calculations, formula (4.59). It occurs that the obtained fit is fairly good for $\alpha = 0.35$. Considering that for $T = 0$ hadrons $\alpha = 0.13$, the obtained value of $\alpha = 0.35$ points to the thermally excited nucleons [4.30].

For relativistic *mesons*, we can define the transversal mass

$$m_t = \sqrt{c^2 p_t^2 + m_0^2}\,, \tag{4.60}$$

where p_t denotes the momentum of particle emitted at $90°$ to the laser beam direction. The invariant cross section for the transversal mass has the form [4.31]

$$\frac{1}{m_t}\frac{dN}{dm_t} = \int_0^{t_{\text{life}}} m_t K\left[1, \frac{m_t}{T_h}\right] dt\,. \tag{4.61}$$

In formula (4.61) $K[1,x]$ denotes the Bessel modified function of the first order, and t_{life} denotes the lifetime of hot *zone* created as the results of thermal excitation.

In Fig. 4.11, we present the comparison of the calculated and measured of the π spectra in heavy ion reactions. In that case, the $T(\boldsymbol{r}, t)$ are the solution of equations (4.49) for initial conditions appropriate for heavy-ion reactions with $\alpha = 0.35$ and $t_{\text{lifetime}} \sim 7$ fm/c.

The nonrelativistic definition of the *heaton* temperature

$$T_h = m_0 \alpha^2 c^2 \qquad (4.62)$$

implies that $T_h < m_0 c^2$ as $\alpha < 1$ for electromagnetic and strong interactions. On the other hand, it must be recognized that in *chromodynamics* [4.32] nucleon-nucleon and presumably quark-quark interactions are energy dependent, i.e. $\alpha^s = \alpha^s(q^2)$ where $q^2 = (p_i - p_j)^2$ is the four-momentum transfer $(p = (\boldsymbol{p}, E))$. In that case, "constant" α is the running constant [4.32]. From the relativistic definition of the *heaton* temperature

$$T_h = m_0 \alpha^2 c^2 (1 - \alpha^2)^{-1/2} \qquad (4.63)$$

we can calculate the α as the function of T_h.

$$\alpha^2 = -\frac{1}{2}\left(\frac{T_h}{m_0 c^2}\right)^2 + \frac{1}{2}\left(\frac{T_h}{m_0 c^2}\right)^2 \left(1 + 4\left(\frac{m_0 c^2}{T_h}\right)^2\right)^{1/2}. \qquad (4.64)$$

In Fig. 4.12, we present the $\alpha^2(T_h)$. It is quite interesting that for $T_h \to \infty$, $\alpha \to 1$ independent of the nature of the interaction (electromagnetic, strong). The line $\alpha = 1$ corresponds to the *photons* and quark heatons [4.7]. Figure 4.12 illustrates the quark "*confinement.*" In all cases, when we intend to transform the gas of electrons, nucleus, or nucleons to massive quark gas $(m_q \neq 0)$, we need the infinite (i.e. impossible) amount of energy $T_h \to \infty$. It can be concluded that the quark *confinement* is the simple result of the special relativity theory (SRT) – with finite amount of energy we cannot accelerate massive particle to velocity $v = c$. It seems that for existing (RHIC) or planned (LHC) accelerators, quarks are out of the hand, of experimenters.

We can make a remark that the experiments with huge accelerators in which the heavy ions (which consists of quarks) are accelerated to very high velocities v, but always $v < c$, resembles the Bertozzi [4.28] experiment in which the electrons are accelerated to energy $E_{\text{kin}} \gg 0.5$ MeV but $v_{\text{electron}} < c$.

As we have not the possibility to observe the massive particle with $v = c$ on the same footing, we can not observe the massive quarks with $v = c$. The serious problem is that no availability of the free massive quarks is the end of reductionism [4.33].

What we need is to find the Rosetta Stone in order to understand the language the Universe is speaking on the subnucleon structure of matter and spacetime. It seems that the ultrahigh energy light emitted by super PW lasers can help us to find the long-awaited Rosetta Stone.

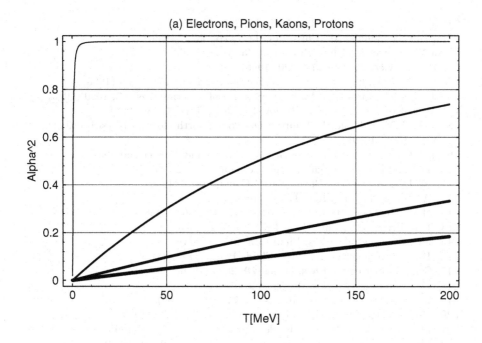

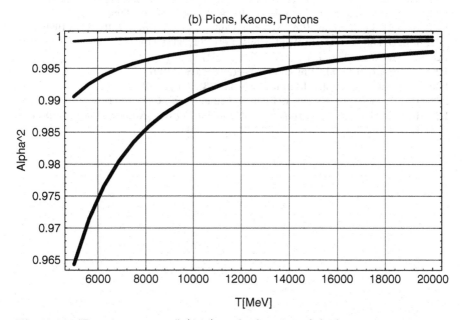

Fig. 4.12. "Running constant" (4.64) as the function of the heaton temperature.

References

4.1. M. Nisoli et al.: Phys. Rev. Lett. **78**, 3575 (1997)

4.2. U. Bockelmann et al.: Phys. Rev. Lett. **76**, 3622 (1996)

4.3. J-Y. Bigot et al.: Phys. Rev. Lett. **75**, 4702 (1995)

4.4. R.F. Service: Science **271**, 920 (1996)

4.5. C.B. Murray, C.R. Kagan, M.G. Bawendi: Science **270**, 1335 (1995)

4.6. J. Marciak-Kozlowska, M. Kozlowski: Int. J. Thermophys. **17**, 1099 (1996)

4.7. M. Kozlowski, J. Marciak-Kozlowska: Lasers Eng. **7**, 13 (1998)

4.8. J.-M. Lévy-Leblond, F. Balibar: *Quantics* (North Holland Physics Publishing Amsterdam 1990)

4.9. J.L. Synge, *The Relativistic Gas* (North Holland Amsterdam 1957)

4.10. D. Strickland, G. Mourou, Opt. Commun. **56**, 219 (1985)

4.11. R.M. More et al.: J. Phys. (Paris) Colloq. **49**, 7 (1988)

4.12. A. Ng. et al.: Phys. Rev. **E51**, 5208 (1995)

4.13. H.M. Milchberg et al.: Phys. Rev. Lett. **61**, 2364 (1988)

4.14. D.F. Price et al.: Phys. Rev. Lett. **75**, 252 (1995)

4.15. J.C. Kieffer et al.: Phys. Fluids **B5**, 2676 (1993)

4.16. D.D. Meyerhofer et al.: Phys. Fluids **B5**, 2584 (1993)

4.17. S.J. Gitomer et al.: Phys. Fluids **29**, 2679 (1986)

4.18. U. Teubner et al.: Phys. Plasmas **2**, 972 (1995)

4.19. G. Guethein et al.: Phys. Rev. Lett. **77**, 1055 (1996)

4.20. N.A. Papadogiannis et al.: Phys. Rev. Lett. **83**, 4289 (1999)

4.21. M. Kozlowski, J. Marciak-Kozlowska: Lasers Eng. **7**, 81 (1998)

4.22. M. Kozlowski, J. Marciak-Kozlowska: Lasers Eng. **9**, 39 (1999)

4.23. D.D. Awschalom, J.M. Kikkawa: Phys. Today **52**, 33 (1999)

4.26. K.W.D. Ledingham, P.A. Norreys: Contemporary Physics **40**, 367 (1999)

4.27. P.L. Shkolnikov et al.: Appl. Phys. Lett. **71**, 3471 (1997)

4.28. W. Bertozzi: Am. J. Phys. **32**, 551 (1964)

4.29. I.G. Bearden et al.: Phys. Rev. Lett. **78**, 2080 (1997)

4.30. M. Kozlowski et al.: Hadronic Journal **23**, 189 (2000)

4.31. R. Hagedorn: La Rivista del Nuovo Cimento vol. **6**, 1 (1983)

4.32. D.H. Perkins, *Introduction to High Energy Physics* (Addison-Wesley 1987)

4.33. R.B. Laughlin, D. Pines, PNAS **97**, 28 (2001)

5

Causal Thermal Phenomena in a Planck Era

5.1 The Time Arrow in a Planck Gas

5.1.1 Quantum Heat Transport in a Planck Era

The enigma of the Planck era, i.e. the event characterized by the Planck time, Planck radius, and Planck mass, is very attractive for speculations. In this section, we discuss the new interpretation of the Planck time. We define Planck gas – a gas of massive particles all with masses equal to the Planck mass $M_P = (\hbar c/G)^{1/2}$, and relaxation for transport process equals the Planck time $\tau_P = (\hbar G/c^5)^{1/2}$. To the description of a thermal transport process in a Planck gas, we apply the quantum heat transport equation (QHT) derived in Chapter 2. The QHT is the specification of the hyperbolic heat conduction equation HHT [5.1, 5.2] to the quantum limit of heat transport, i.e. when the de Broglie wavelength λ_B equals the mean free path, λ.

In the following, we will describe the thermal properties of the Planck gas. To that aim, we use the hyperbolic heat transport equation (HHC) (3.31), (3.32)

$$\frac{\lambda_B}{v_h} \frac{\partial^2 T}{\partial t^2} + \frac{\lambda_B}{\lambda} \frac{\partial T}{\partial t} = \frac{\hbar}{M_P} \nabla^2 T. \tag{5.1}$$

In Eq. (5.1), M_P is the Planck mass, λ_B the de Broglie wavelength, and λ is mean free path for Planck mass. The HHC equation describes the dissipation of the thermal energy induced by a temperature gradient ∇T. Recently, the dissipation processes in the cosmological context (e.g. viscosity) was described in the frame of EIT (extended irreversible thermodynamics) [5.2, 5.3]. With the simple choice for viscous pressure, it is shown that dissipative signals propagate with the light velocity, c [5.2]. Considering that the relaxation time τ is defined as [5.1]

$$\tau = \frac{\hbar}{M_\mathrm{P} v_h^2} \,, \tag{5.2}$$

for thermal wave velocity $v_h = c$, one obtains

$$\tau = \frac{\hbar}{M_\mathrm{P} c^2} = \left(\frac{\hbar G}{c^5}\right)^{1/2} = \tau_\mathrm{P} \,, \tag{5.3}$$

i.e. *the relaxation time is equal to the Planck time* τ_P. The gas of massive particles with masses equal to the Planck mass M_P, and relaxation time for transport processes that equals Planck time τ_P, we will define as the Planck gas.

According to the result of the paper [5.1], we define the quantum of the thermal energy, the *heaton* for the Planck gas, as

$$E_h = \hbar\omega = \frac{\hbar}{\tau_\mathrm{P}} = \left(\frac{\hbar c}{G}\right)^{1/2} c^2 = M_\mathrm{P} c^2,$$

$$E_h = M_\mathrm{P} c^2 = E^{\mathrm{Planck}} = 10^{19}\,\mathrm{GeV}. \tag{5.4}$$

With formula (5.2) and $v_h = c$, we calculate the mean free path, λ, viz.

$$\lambda = v_h \tau_\mathrm{P} = c\tau_\mathrm{P} = c\left(\frac{\hbar G}{c^5}\right)^{1/2} = \left(\frac{\hbar G}{c^3}\right)^{1/2}. \tag{5.5}$$

From formula (5.5), we conclude that mean free path for a Planck gas is equal to the Planck radius. For a Planck mass, we can calculate the de Broglie wavelength

$$\lambda_\mathrm{B} = \frac{\hbar}{M_\mathrm{P} v_h} = \frac{\hbar}{M_\mathrm{P} c} = \left(\frac{G\hbar}{c^3}\right)^{1/2} = \lambda. \tag{5.6}$$

As it is defined in paper [5.1], Eq. (5.6) describes the quantum limit of heat transport. When formulas (5.5) and (5.6) are substituted in Eq. (5.1), we obtain

$$\tau_\mathrm{P}\frac{\partial^2 T}{\partial t^2} + \frac{\partial T}{\partial t} = \frac{\hbar}{M_\mathrm{P}}\nabla^2 T. \tag{5.7}$$

Equation (5.7) is the quantum hyperbolic heat transport equation (QHT) for a Planck gas. It can be written as

$$\frac{\partial^2 T}{\partial t^2} + \left(\frac{c^5}{\hbar G}\right)^{1/2}\frac{\partial T}{\partial t} = c^2\nabla^2 T. \tag{5.8}$$

It is interesting to observe that QHT is the damped wave equation, and gravitation influences the dissipation of the thermal energy. In paper [5.4],

P.G. Bergman discussed the conditions for the thermal equilibrium in the presence of the gravitation. As it was shown in that paper, the thermal equilibrium of spatially extended systems is characterized by the "global" temperature and a "local" temperature that is sensitive to the value of the gravitational potential.

On the other hand, Eq. (5.8) describes the correlated random walk of Planck mass. For mean square displacement of random walkers, we have

$$<x^2> = \frac{2\hbar}{M_P}\left[\frac{t}{\tau_P} - \left(1 - e^{-t/\tau_P}\right)\right].\tag{5.9}$$

From formula (5.6), we conclude that for $t \sim \tau_P$,

$$<x^2> \cong \frac{\hbar}{M_P\tau}t^2\tag{5.10}$$

or

$$<x^2> \cong c^2t^2,\tag{5.11}$$

and we have thermal wave with velocity c. For $t \gg \tau_P$, we have

$$<x^2> \sim \frac{2\hbar\tau}{M_P}\left(\frac{t}{\tau_P} - 1\right) = \frac{2\hbar}{M_P}t = 2D^{\mathrm{Planck}}t,\tag{5.12}$$

where

$$D^{\mathrm{Planck}} = \frac{\hbar}{M_P} = \left(\frac{\hbar G}{c}\right)^{1/2}\tag{5.13}$$

denotes the diffusion coefficient for a Planck mass.

We can conclude that, for time period of the order the Planck time, QHT describes the propagation of a thermal wave with velocity equal c and, for a time period much longer than τ_P, QHT describes the diffusion process with diffusion coefficient dependent on the gravitation constant G.

The quantum hyperbolic heat equation (5.7), as a hyperbolic equation sheds light on the time arrow in a Planck gas. When QHT is written in the equivalent form

$$\tau_P\frac{\partial^2 T}{\partial t^2} + \frac{\partial T}{\partial t} = \left(\frac{\hbar G}{c}\right)^{1/2}\nabla^2 T,\tag{5.14}$$

then, for a time period shorter than τ_P, we have preserved time reversal for thermal processes, viz.,

$$\frac{1}{c^2}\frac{\partial^2 T}{\partial t^2} = \nabla^2 T,\tag{5.15}$$

and for $t \gg \tau_P$,

$$\frac{\partial T}{\partial t} = \left(\frac{\hbar G}{c}\right)^{1/2} \nabla^2 T \tag{5.16}$$

the time reversal symmetry is broken.

These new properties of QHT open up new possibilities for the interpretation of the Planck time. Before τ_P, thermal processes in Planck gas are symmetrical in time. After τ_p the time symmetry is broken. Moreover, it seems that gravitation is activated after τ_p, and it creates an arrow of time (formula (5.16)).

5.1.2 The Smearing Out of the Thermal Initial Conditions Created in a Planck Era

In paper [5.5], the QHT for a Planck gas was discussed. On time scales of Planck time, black holes of the Planck mass spontaneously come into existence. Via the process of Hawking radiation, the black hole can then evaporate back into energy. The characteristic time scale for this to occur happens to be approximately equal to Planck time. Thus, the Universe at $t_P = 10^{-43}$ s in age was filled with a Planck gas.

In the subsequent paper, we develop the generalized quantum heat transport equation for Planck gas, which includes the potential energy term. The condition for conserving the shape of the thermal wave created at the Planck time is developed and investigated.

For a long time, the analogy between the Schrödinger equation and the diffusion equation was recognized [5.6]. Let us consider, for the moment, the parabolic heat transport equation for a Planck gas, i.e. (5.16),

$$\frac{\partial T}{\partial t} = \frac{\hbar}{M_P} \nabla^2 T. \tag{5.17}$$

When the real time $t \to \frac{it}{2}$ and $T \to \Psi$, (5.17) has the form of the free Schrödinger equation

$$i\hbar \frac{\partial \Psi}{\partial t} = -\frac{\hbar^2}{2M_P} \nabla^2 \Psi. \tag{5.18}$$

The complete Schrödinger equation has the form

$$i\hbar \frac{\partial \Psi}{\partial t} = -\frac{\hbar^2}{2M_P} \nabla^2 \Psi + V \Psi, \tag{5.19}$$

where V denotes the potential energy. When we go back to real time $t \to -2it$ and $\Psi \to T$, the new parabolic quantum heat transport is obtained

$$\frac{\partial T}{\partial t} = \frac{\hbar}{M_P} \nabla^2 T - \frac{2V}{\hbar} T \,. \tag{5.20}$$

Equation (5.20) describes the quantum heat transport in a Planck gas for $\Delta t > t_P$. For heat transport in the period $\Delta t < t_P$, one obtains the generalized hyperbolic heat transport equation [5.5] with potential term added

$$t_P \frac{\partial^2 T}{\partial t^2} + \frac{\partial T}{\partial t} = \frac{\hbar}{M_P} \nabla^2 T - \frac{2V}{\hbar} T \,. \tag{5.21}$$

Considering that $t_P = \hbar/M_P c^2$ [5.5], (5.21) can be written as

$$\frac{1}{c^2} \frac{\partial^2 T}{\partial t^2} + \frac{M_P}{\hbar} \frac{\partial T}{\partial t} + \frac{2V M_P}{\hbar^2} T = \nabla^2 T \,, \tag{5.22}$$

where c denotes light velocity in vacuum.

In the following, we consider the one-dimensional heat transport phenomena with constant potential energy $V = V_0$

$$\frac{1}{c^2} \frac{\partial^2 T}{\partial t^2} + \frac{M_P}{\hbar} \frac{\partial T}{\partial t} + \frac{2V_0 M_P}{\hbar^2} T = \frac{\partial^2 T}{\partial x^2} \,. \tag{5.23}$$

For quantum heat transport equation (5.23), we seek a solution in the form

$$T(x,t) = e^{-t/t_P} u(x,t) \,. \tag{5.24}$$

After substituting (5.24) in (5.23), one obtains

$$\frac{\partial^2 u(x,t)}{\partial t^2} = c^2 \frac{\partial^2 u}{\partial x^2} + b^2 u(x,t) \,, \tag{5.25}$$

where

$$b = \sqrt{\left(\frac{M_P c^2}{2\hbar} \right)^2 - \frac{2V_0 M_P}{\hbar^2} c^2} \,. \tag{5.26}$$

The general solution of (5.26) for Cauchy initial conditions

$$u(x,0) = f(x), \qquad \left. \frac{\partial u(x,t)}{\partial t} \right|_{t=0} = F(x) \,, \tag{5.27}$$

has the form [5.7]

$$u(x,t) = \frac{f(x-ct) + f(x+ct)}{2} + \frac{1}{2} \int_{x-ct}^{x+ct} \Phi(x,t,z) \mathrm{d}z \,, \tag{5.28}$$

where

$$\Phi(x,t,z) = \frac{1}{c} F(z) J_0 \left(\frac{b}{c} \sqrt{(z-x)^2 - c^2 t^2} \right)$$

$$+ b t f(z) \frac{J_0' \left(\frac{b}{c} \sqrt{(z-x)^2 - c^2 t^2} \right)}{\sqrt{(z-x)^2 - c^2 t^2}} \tag{5.29}$$

and $J_0(z)$ denotes the Bessel function of the first kind. Considering formulas (5.24–5.27), the solution of (5.23) describes the propagation of the initial state $f(x)$ of the Planck gas as the thermal wave with velocity c. It is quite interesting to formulate the condition at which these waves propagate without the distortion, i.e. conserving their shapes. The conditions for this to happen can be formulated as

$$b = \sqrt{\left(\frac{M_P c^2}{2\hbar}\right)^2 - \frac{2V_0 M_P}{\hbar^2} c^2} = 0.$$ (5.30)

When (5.30) holds, (5.25) assumes the form

$$\frac{\partial^2 u(x,t)}{\partial t^2} = c^2 \frac{\partial^2 u}{\partial x^2}.$$ (5.31)

Equation (5.31) is the wave equation with the solution (for Cauchy initial conditions (5.27))

$$u(x,t) = \frac{f(x - ct) + f(x + ct)}{2} + \frac{1}{2c} \int_{x-ct}^{x+ct} F(z)dz.$$ (5.32)

Equation (5.30), the distortionless condition, can be written as

$$V_0 t_P = \frac{\hbar}{8} \sim \hbar.$$ (5.33)

We can conclude that in the presence of the potential V_0, one can "observe" the undisturbed quantum thermal wave (created at $t = 0$) only when the *Heisenberg uncertainty* relation (5.33) is fulfilled.

On combining (5.24) and (5.32), the complete solution of (5.23) (for $b = 0$) can be written as

$$T(x,t) = e^{-t/2t_P} \left[\frac{f(x - ct) + f(x + ct)}{2} + \frac{1}{2c} \int_{x-ct}^{x+ct} F(z)dz \right].$$ (5.34)

One can say that the formula (5.34) is a very pessimistic one, because the initial conditions (which operate at the Beginning) are smeared out over a time scale of the order the Planck time.[1]

---

[1] Information loss at the Planck scale may also shed further light on the origin of the gauge theories. Compare: Gerard 't Hooft Determinism Beneath Quantum Mechanics, In: A. Elitzur, S. Doler, N. Kolenda eds. *Quo Vadis Quantum Mechanics* (Springer 2005) p 99.

5.2 Klein-Gordon Thermal Equation for a Planck Gas

5.2.1 Planck Wall Potential

As was shown in paper [5.5], the thermal properties of the Planck gas can be described by hyperbolic quantum heat transport equation, viz.,

$$\frac{1}{c^2}\frac{\partial^2 T}{\partial t^2} + \frac{M_{\mathrm{P}}}{\hbar}\frac{\partial T}{\partial t} + \frac{2VM_{\mathrm{P}}}{\hbar^2}T = \nabla^2 T. \tag{5.35}$$

In Eq. (5.35), t_{P} denotes Planck time, M_{P} is the Planck mass, and V denotes the potential energy.

For the uniform Universe, it is possible to study only one-dimensional heat transport phenomena. In the following, we will consider the thermal properties of a Planck gas for constant potential $V = V_0$. In that case, the one-dimensional quantum heat transport equation has the form

$$\frac{1}{c^2}\frac{\partial^2 T}{\partial t^2} + \frac{M_{\mathrm{P}}}{\hbar}\frac{\partial T}{\partial t} + \frac{2V_0 M_{\mathrm{P}}}{\hbar^2}T = \frac{\partial^2 T}{\partial x^2}, \tag{5.36}$$

where formula for $t_{\mathrm{P}} = \hbar/M_{\mathrm{P}}c^2$ was used [5.5]. In (5.36), c denotes the light velocity. As $c \neq \infty$, we cannot omit the second derivative term and consider only Fokker-Planck equation

$$\frac{M_{\mathrm{P}}}{\hbar}\frac{\partial T}{\partial t} + \frac{2V_0 M_{\mathrm{P}}}{\hbar^2}T = \frac{\partial^2 T}{\partial x^2} \tag{5.37}$$

for heat diffusion in the potential energy V_0, or free heat diffusion

$$\frac{\partial T}{\partial t} = \frac{\hbar}{M_{\mathrm{P}}}\frac{\partial^2 T}{\partial x^2}. \tag{5.38}$$

It occurs that only if we retain the second derivative term, we have the chance to study the conditions in the Beginning.

Some implications of the forward and backward properties of the parabolic heat diffusion equation were beautifully described by J.C. Maxwell [5.7]:

"Sir William Thompson has shown in a paper published in the Cambridge and Dublin Mathematical Journal in 1844 how to deduce, in certain cases the thermal state of a body in past time from its observed conditions at present. If the present distribution of temperature is such that it may be expressed in a finite series of harmonics, the distribution of temperature at any previous time may be calculated but if (as in generally case) the series of harmonics is infinite, than the temperature can be calculated only when this series is convergent. For present and future time it is always convergent, but for past

time it becomes ultimately divergent when the time is taken at a sufficiently remote epoch. The negative value of t for which the series becomes ultimately divergent, indicates a certain date in past time such that the present state of things cannot be deduced from any distribution of temperature occurring previously to the date, and becoming diffused by ordinary conduction. Some other event besides ordinary conduction must have occurred since that date in order to produce the present stage of things."

As can be easily seen, the second derivative term in (5.35) carriers the memory of the initial state that occurred at time $t = 0$. If we pass with $c \to \infty$, we lose the possibility of studying the influence of the initial conditions at the present epoch as it is explained above by J.C. Maxwell. It means that by limiting procedure $c \to \infty$, *we cut off the memory of the Universe.*

For hyperbolic quantum heat transport, (5.36), we seek a solution of the form

$$T(x,t) = e^{-t/2t_P} u(x,t) .\tag{5.39}$$

After substitution of (5.39) in (5.36), one obtains

$$\frac{1}{c^2}\frac{\partial^2 u}{\partial t^2} - \frac{\partial^2 u}{\partial x^2} + qu = 0 ,\tag{5.40}$$

where

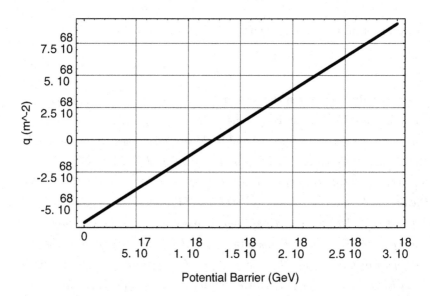

Fig. 5.1. Parameter q (formula 5.36) as the function of the barrier height (GeV).

$$q = \frac{2V_0 M_P}{\hbar^2} - \left(\frac{M_P c}{2\hbar}\right)^2. \tag{5.41}$$

In the following, we shall consider positive values of V_0, $V_0 \geq 0$, i.e. as well as the potential barriers and steps.

The structure of (5.40) depends on the sign of the parameter q. Let us define the Planck wall potential, i.e. potential for which $q = 0$. From (5.41), one obtains

$$V_P = \frac{\hbar}{8t_P} = 1.25\ 10^{18}\ \text{GeV}, \tag{5.42}$$

where t_P is a Planck time.

In Fig. 5.1, the parameter q is calculated as the function of V_0. For $q < 0$, i.e. when $V_0 < V_P$, (5.40) is *the modified telegraph equation* (MTE) [5.3]. For the Cauchy initial condition

$$u(x,0) = f(x), \qquad \frac{\partial u(x,0)}{\partial t} = g(x), \tag{5.43}$$

and the solution of (5.39) has the form [5.3]

$$u(x,t) = \frac{f(x - ct) + f(x + vt)}{2}$$

$$+ \frac{1}{2c} \int_{x-ct}^{x+ct} g(\zeta) I_0 \left[\sqrt{-q(c^2 t^2 - (x - \zeta)^2)}\right] d\zeta$$

$$+ \frac{(c\sqrt{-q})t}{2} \int_{x-ct}^{x+ct} f(\zeta) \frac{I_1 \left[\sqrt{-q(c^2 t^2 - (x - \zeta)^2)}\right]}{\sqrt{c^2 t^2 - (x - \zeta)^2}} d\zeta. \tag{5.44}$$

In equation (5.44), I_0, I_1 denotes the Bessel modified function of the zero and one, respectively. When $q > 0$, i.e. for $V_0 > V_P$, (5.40) is reduced to *the Klein-Gordon equation* (K-GE), well-known from its application in elementary particle and nuclear physics. For the Cauchy initial condition (5.43), the solution of K-GE can be written as [5.3]

$$u(x,t) = \frac{f(x - ct) + f(x + ct)}{2}$$

$$+ \frac{1}{2c} \int_{x-ct}^{x+ct} g(\zeta) J_0 \left[\sqrt{q(c^2 t^2 - (x - \zeta)^2)}\right] d\zeta \tag{5.45}$$

$$- \frac{(c\sqrt{q})t}{2} \int_{x-ct}^{x+ct} \frac{J_1 \left[\sqrt{q(c^2 t^2 - (x - \zeta)^2)}\right]}{\sqrt{c^2 t^2 - (x - \zeta)^2}} d\zeta.$$

The case for $q = 0$ was discussed in paper [5.5] and described the distortionless quantum thermal waves.

Both solutions (5.44) and (5.45) exhibit the domains of dependence and influence on the modified telegraph equation and Klein-Gordon equation. These domains, which characterize the maximum speed, c, at which the thermal disturbance travels are determined by the principal terms of the given equation (i.e. the second derivative terms), do not depend on the lower order terms. It can be concluded that these equations and the wave equation have identical domains of dependence and influence. Both solutions (5.44) and (5.45) represent the distorted thermal waves in the field of potential barrier or steps V.

5.2.2 Radius, Velocity, and Acceleration of the Spacetime

In recent years, the growing interest for the source of Universe expansion is observed. After the work of supernova-detecting groups, the consensus for the acceleration of the moving of the spacetime is established [5.8, 5.9].

In the subsequent, we follow the idea of the repulsive gravity as the source of the spacetime expansion. We will study the influence of the repulsive gravity $(G < 0)$ on the temperature field in the universe. To that aim, we will apply the quantum hyperbolic heat transfer equation (QHT) formulated in our earlier papers [5.5, 5.10].

When substitution $G \to -G$ is performed in QHT, the Schrödinger type equation is obtained for the temperature field. In this section, the solution of QHT will be obtained. The resulting temperature is a complex function of space and time. We argue that because of the anthropic limitation of the observers, it is quite reasonable to assume $\mathrm{Im}\, T = 0$. From this anthropic condition, the discretization of the space radius $R = [(4N\pi + 3\pi)L_{\mathrm{P}}]^{1/2}(ct)^{1/2}$, velocity of expansion $v = (\pi/4)^{1/2}((N + \frac{3}{4})/M)^{1/2}c$, and acceleration of expansion $a = -\frac{1}{2}(\pi/4)^{1/2}((N + \frac{3}{4})^{1/2}/M^{3/2})(c^7/(\hbar G))^{1/2}$ are obtained.

In papers [5.5, 5.10], the quantum heat transport equation in a Planck era was formulated:

$$\tau \frac{\partial^2 T}{\partial t^2} + \frac{\partial T}{\partial t} = \frac{\hbar}{M_{\mathrm{P}}} \nabla^2 T. \tag{5.46}$$

In equation (5.46), $\tau = ((\hbar G)/c^5)^{1/2}$ is the relaxation time, $M_{\mathrm{P}} = ((\hbar c)/G)^{1/2}$ is the mass of the Planck particle, $\hbar, c$ are the Planck constant and light velocity, respectively, and G is the gravitational constant. The crucial role played by gravity (represented by G in formula (5.46)) in a Planck era was investigated in paper [5.10].

For a long time the question whether or not the fundamental constant of Nature G varies with time has been a question of considerable interest. Since

P.A.M. Dirac [5.11] suggested that the gravitational force may be weakening with the expansion of the Universe, a variable G is expected in theories such as the Brans-Dicke scalar-tensor theory and its extension [5.12, 5.13]. Recently, the problem of the varying G received renewed attention in the context of extended inflation cosmology [5.14].

It is now known that the spin of a field (electromagnetic, gravity) is related to the nature of the force: fields with odd-integer spins can produce both attractive and repulsive forces; those with even-integer spins such as scalar and tensor fields produce a purely attractive force. Maxwell's electrodynamics for instance can be described as a spin one field. The force from this field is attractive between oppositely charged particles and repulsive between similarly charged particles.

The integer spin particles in gravity theory are like the graviton, mediators of forces and would generate the new effects. Both the graviscalar and the graviphoton are expected to have the rest mass and so their range will be finite rather than infinite. Moreover, the graviscalar will produce only attraction, whereas the graviphoton effect will depend on whether the interacting particles are alike or different. Between matter and matter (or antimatter and antimatter), the graviphoton will produce repulsion.

The existence of repulsive gravity forces can to some extent explain the early expansion of the Universe [5.11].

Now we will describe the influence of the repulsion gravity on the quantum thermal processes in the universe. To that aim, we put in equation (5.46) $G \to -G$. In that case, the new equation is obtained, viz.

$$i\hbar \frac{\partial T}{\partial t} = \left(\frac{\hbar^3 |G|}{c^5} \right)^{1/2} \frac{\partial^2 T}{\partial t^2} - \left(\frac{\hbar^3 |G|}{c} \right)^{1/2} \nabla^2 T. \tag{5.47}$$

For the investigation of the structure of (5.47) we put:

$$\frac{\hbar^2}{2m} = \left(\frac{\hbar^3 |G|}{c} \right)^{1/2} \tag{5.48}$$

and obtain

$$m = \frac{1}{2} M_P$$

with new form of the equation (5.47)

$$i\hbar \frac{\partial T}{\partial t} = \left(\frac{\hbar^3 |G|}{c^5} \right)^{1/2} \frac{\partial^2 T}{\partial t^2} - \frac{\hbar^2}{2m} \nabla^2 T. \tag{5.49}$$

Equation (5.49) is the quantum telegraph equation discussed in paper [5.10]. To clarify the physical nature of the solution of (5.49), we will discuss the diffusion approximation, i.e. we omit the second time derivative in (5.49) and obtain

$$i\hbar\frac{\partial T}{\partial t} = -\frac{\hbar^2}{2m}\nabla^2 T. \tag{5.50}$$

Equation (5.50) is the Schrödinger type equation for the temperature field in a universe with $G < 0$.

Both equation (5.50) and diffusion equation:

$$\frac{\partial T}{\partial t} = \frac{\hbar^2}{2m}\nabla^2 T \tag{5.51}$$

are parabolic and require the same boundary and initial conditions in order to be "well posed."

The diffusion equation (5.51) has the propagator [5.16]:

$$T_D(\boldsymbol{R}, \Theta) = \frac{1}{(4\pi D\Theta)^{3/2}}\exp\left[-\frac{R^2}{2\pi\hbar\Theta}\right], \tag{5.52}$$

where

$$\boldsymbol{R} = \boldsymbol{r} - \boldsymbol{r}', \qquad \Theta = t - t'.$$

For equation (5.50), the propagator is

$$T_s(\boldsymbol{R}, \Theta) = \left(\frac{M_P}{2\pi\hbar}\right)^{3/2}\exp\left[-\frac{3\pi i}{4}\right]\cdot\exp\left[\frac{iM_P R^2}{2\pi\hbar\Theta}\right] \tag{5.53}$$

with initial condition $T_s(\boldsymbol{R}, 0) = \delta(\boldsymbol{R})$.

In equation (5.53), $T_s(\boldsymbol{R}, \Theta)$ is the complex function of $\boldsymbol{R}$ and Θ. For anthropic observers only the real part of T is detectable, so in our description of universe we put:

$$\mathrm{Im}T(\boldsymbol{R}, \Theta) = 0. \tag{5.54}$$

The condition (5.54) can be written as (bearing in mind formula (5.53)):

$$\sin\left[-\frac{3\pi}{4} + \left(\frac{R}{L_P}\right)^2\frac{1}{4\widetilde{\Theta}}\right] = 0, \tag{5.55}$$

where $L_P = \tau_P c$ and $\widetilde{\Theta} = \Theta/\tau_P$. Formula (5.55) describes the discretization of R

$$R_N = [(4N\pi + 3\pi)L_P]^{1/2}(tc)^{1/2}, \tag{5.56}$$

$$N = 0, 1, 2, 3\ldots$$

In fact from formula (5.56), the Hubble law can be derived

$$\frac{\dot{R}_N}{R_N} = H = \frac{1}{2t}, \qquad \text{independent of } N. \tag{5.57}$$

In the subsequent, we will consider R (5.56), as the spacetime radius of the N-universe with "atomic unit" of space L_P.

It is well-known that idea of discrete structure of time can be applied to the "flow" of time. The idea that time has "atomic" structure or is not infinitely divisible has only recently come to the fore as a daring and sophisticated hypothetical concomitant of recent investigations in elementary particle physics and astrophysics. Yet in the Middle Ages, the atomicity of time was maintained by various thinkers, notably by Maimonides [5.17]. In the most celebrated of his works, *The Guide for Perplexed*, he wrote: *Time is composed of time-atoms, i.e. of many parts, which on account of their short duration cannot be divided.* The theory of Maimonides was also held by Descartes [5.18].

The shortest unit of time, atom of time, is named *chronon* [5.19]. Modern speculations concerning the *chronon* have often be related to the idea that the smallest natural length is L_P. If this is divided by velocity of light, it gives the Planck time $\tau_P = 10^{-43}$ s, i.e. *the chronon is equal, τ_P*. In that case, the time t can be defined as

$$t = M\tau_P, \qquad M = 0, 1, 2, \ldots \tag{5.58}$$

Considering formulae (5.53) and (5.58), the spacetime radius can be written as

Table 5.1. Radius, Velocity, and Acceleration for N, M-universes

N, M	R(m)	v(m/s)	a(m/s^2)
10^{20}	$1.77\ 10^{-15}$	$2.66\ 10^8$	$-1.32\ 10^{31}$
10^{60}	$1.77\ 10^{25\,(*)}$	$2.66\ 10^{8\,(*)}$	$-1.32\ 10^{-10\,(**)}$
10^{80}	$1.77\ 10^{45}$	$2.66\ 10^8$	$-1.32\ t10^{-29}$

(*)D.N. Spergel et al. [5.22];
(**)J.D. Anderson et al. [5.23] Radio metric data from Pionier 10/11, Galileo and Ulysses Data indicate an apparent anomalous, constant acceleration acting on the spacecraft with a magnitude $\sim 8.5\ 10^{-10}$ m/s^2.

$$R(M, N) = (\pi)^{1/2} M^{1/2} \left(N + \frac{3}{4} \right)^{1/2} L_\text{P}, \qquad M, N = 0, 1, 2, 3, \ldots \quad (5.59)$$

Formula (5.59) describes the discrete structure of spacetime. As the $R(M, N)$ is time dependent, we can calculate the velocity, $v = dR/dt$, i.e. the velocity of the expansion of spacetime

$$v = \left(\frac{\pi}{4} \right)^{1/2} \left(\frac{N + 3/4}{M} \right)^{1/2} c, \qquad (5.60)$$

where c is the light velocity. We define the acceleration of the expansion of the spacetime

$$a = \frac{dv}{dt} = -\frac{1}{2} \left(\frac{\pi}{4} \right)^{1/2} \frac{(N + 3/4)^{1/2}}{M^{3/2}} \frac{c}{\tau_\text{P}}. \qquad (5.61)$$

Considering formula (5.61), it is quite natural to define Planck acceleration:

$$A_\text{P} = \frac{c}{\tau_\text{P}} = \left(\frac{c^7}{\hbar G} \right)^{1/2} = 10^{51} \text{ m/s}^2 \qquad (5.62)$$

and formula (5.61) can be written as

$$a = -\frac{1}{2} \left(\frac{\pi}{4} \right)^{1/2} \frac{(N + 3/4)^{1/2}}{M^{3/2}} \left(\frac{c^7}{\hbar G} \right)^{1/2}. \qquad (5.63)$$

In Table 5.1, the numerical values for R, v, and a are presented. It is quite interesting that for $N, M \to \infty$, the expansion velocity $v < c$ is in complete accord with relativistic description. Moreover for $N, M \gg 1$, the v is relatively constant $v \sim 0.88c$. From formulae (5.56) and (5.60), the Hubble parameter H and the age of our Universe can be calculated

$$v = HR, \qquad H = \frac{1}{2M\tau_\text{P}} = 5 \ 10^{-18} \ 1/s,$$

$$T = 2M\tau_\text{P} = 2 \ 10^{17}\text{s} \sim 10^{10} \text{ years}, \qquad (5.64)$$

which is in quite good agreement with recent measurement [5.21, 5.22, 5.23].

In Figs. 5.2(a) and (b) the velocity and acceleration as the function of $L(L_\text{P})$ and $T(T_\text{P})$ are presented, and in Figs. 5.3(a) and (b), the presented day radiuses for $N, M = 10^{60}$ are presented. In this section following the QHT, the discrete structure of the spacetime is investigated. Assuming anthropic condition $\text{Im}T(r, t) = 0$, the discretization of space-time is evaluated. The formulae for discrete radius $R(N, M)$, velocity $v(N, M)$ and acceleration are obtained. It is shown that numerical values $R(N, M) = 10^{60} L_\text{P}$, $v(N, M) = 0.88c$, and $a(N, M) = 4.43 \cdot 10^{-60} A_\text{P}$ for $N = M = 10^{60}$ are in good agreement with the observational data of our Universe.

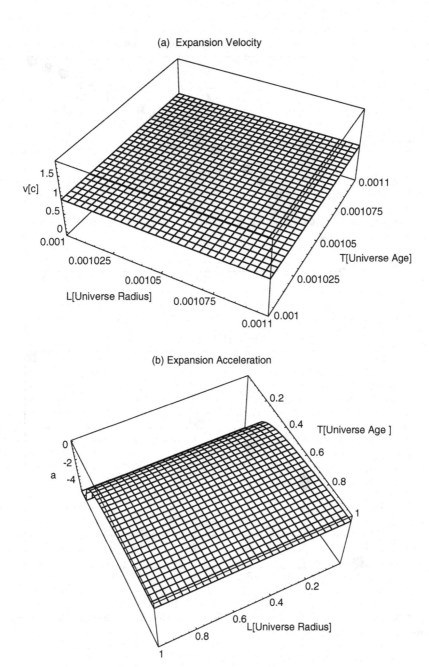

Fig. 5.2. (a) Velocity of the expansion and (b) acceleration of the expansion for spacetime of the N, M universes.

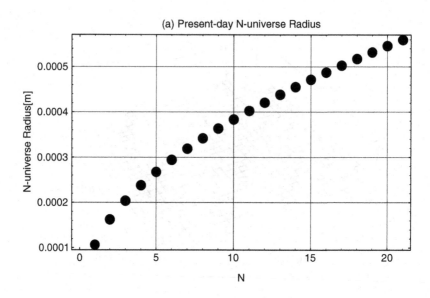

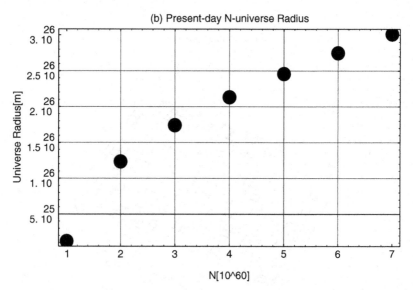

Fig. 5.3. (a) Present-day N-universe radius for small $N \leq 100$. (b) Present-day radius for N-universe $N \sim 10^{60}$.

5.2.3 Stability of Matter in the Accelerating Spacetime

In the seminal paper [5.24], F. Calogero described the cosmic origin of quantization. In paper [5.24], the tremor of the cosmic particles is the origin of the quantization, and the characteristic acceleration of these particles $a \sim 10^{-10}$ m/s^2 was calculated. In our earlier paper [5.25], the same value of the acceleration was obtained and compared with the experimental value of the measured space-time acceleration. In this section, we define the cosmic force *Planck* force, $F_{\text{Planck}} = M_P a_{\text{Planck}}$ ($a_{\text{Planck}} \sim a$), and study the history of Planck force as the function of the age of the Universe.

Masses introduce a curvature in spacetime, light and matter are forced to move according to spacetime metric. Because all the matter is in motion, the geometry of space is constantly changing. A. Einstein relates the curvature of space to the mass/energy density:

$$G = k T . \tag{5.65}$$

G is the Einstein curvature tensor and T the stress-energy tensor. The proportionality factor k follows by comparison with Newton's theory of gravity: $k = G/c^4$ where G is the Newton's gravity constant and c is the vacuum velocity of light; it amounts to about 2.10^{-43} N^{-1}, expressing the *rigidity* of spacetime.

In paper [5.25], the model for the acceleration of spacetime was developed. Prescribing the $-G$ for spacetime and $+G$ for matter, the acceleration of spacetime was obtained:

$$a_{\text{Planck}} = -\frac{1}{2} \left(\frac{\pi}{4}\right)^{1/2} \frac{(N + \frac{3}{4})^{1/2}}{M^{3/2}} A_P , \tag{5.66}$$

where A_P, *Planck* acceleration equals, viz.:

$$A_P = \left(\frac{c^7}{\hbar G}\right)^{1/2} = \frac{c}{\tau_P} \cong 10^{51} \text{m/s}^2 . \tag{5.67}$$

As was shown in paper [5.25], the a_{Planck} for $N = M = 10^{60}$ is of the order the acceleration detected by Pioneer spacecrafts [5.26].

Considering A_P, it is quite natural to define the *Planck* force F_{Planck},

$$F_{\text{Planck}} = M_P A_P = \frac{c^4}{G} = k^{-1} , \tag{5.68}$$

where

$$M_{\rm P} = \left(\frac{\hbar c}{G}\right)^{1/2}.$$

From formula (5.68), we conclude that $F_{\rm Planck}^{-1}$ = rigidity of the spacetime. The *Planck* force, $F_{\rm Planck} = c^4/G = 1.2 \; 10^{44}$ N, can be written in units that characterize the microspace-time, i.e. GeV and fm.

In these units

$$k^{-1} = F_{\rm Planck} = 7.6 \; 10^{38} \text{ GeV/fm}.$$

As was shown in paper [5.25], the present value of *Planck* force equals

$$F_{\rm Planck}^{\rm Now}(N = M = 10^{60}) \cong -\frac{1}{2}\left(\frac{\pi}{4}\right)^{1/2} 10^{-60}\frac{c^4}{G} = -10^{-22}\frac{\text{GeV}}{\text{fm}}. \qquad (5.69)$$

In paper [5.5], the *Planck* time $\tau_{\rm P}$ was defined as the relaxation time for space-time

$$\tau_{\rm P} = \frac{\hbar}{M_{\rm P}c^2}. \qquad (5.70)$$

Considering formulae (5.68) and (5.70), $F_{\rm Planck}$ can be written as

$$F_{\rm Planck} = \frac{M_{\rm P}c}{\tau_{\rm P}}, \qquad (5.71)$$

where c is the velocity for gravitation propagation. In paper [5.1], the velocities and relaxation times for thermal energy propagation in atomic and nuclear matter were calculated:

$$v_{\rm atomic} = \alpha_{em}c, \qquad (5.72)$$

$$v_{\rm nuclear} = \alpha_s c,$$

where $\alpha_{em} = {\rm e}^2/(\hbar c) = 1/137$, $\alpha_s = 0.15$. In the subsequent, we define atomic and nuclear accelerations:

$$a_{\rm atomic} = \frac{\alpha_{em}c}{\tau_{\rm atomic}}, \qquad (5.73)$$

$$a_{\rm nuclear} = \frac{\alpha_s c}{\tau_{\rm nuclear}}.$$

Considering that $\tau_{\rm atomic} = \hbar/(m_e\alpha_{em}^2c^2)$, $\tau_{\rm nuclear} = \hbar/(m_N\alpha_s^2c^2)$, one obtains from formula (5.73)

$$a_{\rm atomic} = \frac{m_e c^3 \alpha_{em}^3}{\hbar}, \qquad (5.74)$$

$$a_{\rm nuclear} = \frac{m_N c^3 \alpha_s^3}{\hbar}.$$

We define, analogously to *Planck* force the new forces: F_{Bohr}, F_{Yukawa}

$$F_{\text{Bohr}} = m_e a_{\text{atomic}} = \frac{(m_e c^2)^2}{\hbar c} \alpha_{em}^3 = 5 \; 10^{-13} \; \frac{\text{GeV}}{\text{fm}} \; , \tag{5.75}$$

$$F_{\text{Yukawa}} = m_N a_{\text{nuclear}} = \frac{(m_N c^2)^2}{\hbar c} \alpha_s^3 = 1.6 \; 10^{-2} \; \frac{\text{GeV}}{\text{fm}} \; .$$

Comparing formulae (5.69) and (5.75), we conclude that gradients of *Bohr* and *Yukawa* forces are much larger than $F_{\text{Planck}}^{\text{Now}}$, i.e.:

$$\frac{F_{\text{Bohr}}}{F_{\text{Planck}}^{\text{Now}}} = \frac{5.10^{-13}}{10^{-22}} \sim 10^9 , \tag{5.76}$$

$$\frac{F_{\text{Yukawa}}}{F_{\text{Planck}}^{\text{Now}}} = \frac{10^{-2}}{10^{-22}} \sim 10^{20} .$$

The formulae (5.76) guarantee present-day stability of matter on the nuclear and atomic levels.

As the time dependence of F_{Bohr} and F_{Yukawa} are not well established, in the subsequent we will assume that α_s and α_{em} do not dependent on time. Considering formulae (5.72) and (5.75), we obtain

$$\frac{F_{\text{Yukawa}}}{F_{\text{Planck}}} = \frac{1}{(\frac{\pi}{4})^{1/2}} \frac{(m_N c^2)^2}{M_P c^2} \frac{\alpha_s^3}{\hbar} T \; , \tag{5.77}$$

$$\frac{F_{\text{Bohr}}}{F_{\text{Planck}}} = \frac{1}{(\frac{\pi}{4})^{1/2}} \frac{(m_e c^2)^2}{M_P c^2} \frac{\alpha_{em}}{\hbar} T \; . \tag{5.78}$$

As can be realized from formulae (5.77), (5.78) in the past, $F_{\text{Planck}} \sim F_{\text{Yukawa}}$ (for $T = 0.002$ s) and $F_{\text{Planck}} \sim F_{\text{Bohr}}$ (for $T \sim 10^8$ s), T = age of universe. The calculated ages define the limits for instability of the nuclei and atoms.

5.2.4 The *Planck*, *Yukawa*, and *Bohr* Particles

In 1900, M. Planck [5.27] introduced the notion of the universal mass, later on called the *Planck* mass

$$M_P = \left(\frac{\hbar c}{G} \right)^{1/2} , \qquad F_{\text{Planck}} = \frac{M_P c}{\tau_P} \; . \tag{5.79}$$

Considering the definition of the *Yukawa* force (5.75)

$$F_{\text{Yukawa}} = \frac{m_N v_N}{\tau_N} = \frac{m_N \alpha_{\text{strong}} c}{\tau_N} \; , \tag{5.80}$$

the formula (5.80) can be written as:

$$F_{\text{Yukawa}} = \frac{m_{\text{Yukawa}}c}{\tau_N} , \tag{5.81}$$

where

$$m_{\text{Yukawa}} = m_N \alpha_{\text{strong}} \cong 147 \frac{\text{MeV}}{c^2} \sim m_\pi . \tag{5.82}$$

From the definition of the *Yukawa* force, we deduced the mass of the particle that mediates the strong interaction – pion mass postulated by Yukawa in [5.28].

Accordingly for *Bohr* force:

$$F_{\text{Bohr}} = \frac{m_e v}{\tau_{\text{Bohr}}} = \frac{m_e \alpha_{em} c}{\tau_{\text{Bohr}}} = \frac{m_{\text{Bohr}} c}{\tau_{\text{Bohr}}} , \tag{5.83}$$

$$m_{\text{Bohr}} = m_e \alpha_{em} = \frac{3.7}{c^2} \text{keV} . \tag{5.84}$$

For the *Bohr* particle, the range of interaction is

$$r_{\text{Bohr}} = \frac{\hbar}{m_{\text{Bohr}} c} \sim 0.1 \text{ nm} , \tag{5.85}$$

which is of the order atomic radius.

Considering the electromagnetic origin of the mass of the *Bohr* particle, the planned sources of hard electromagnetic field, e.g. free electron laser (FEL) at TESLA accelerator (DESY) [5.29], are best suited to the investigation of the properties of the *Bohr* particles.

5.2.5 Possible Interpretation of F_{Planck}, F_{Yukawa}, and F_{Bohr}

In an important work, published in 1951 J. Schwinger [5.30] demonstrated that in the background of a static uniform electric field, the QED vacuum is unstable and decayed with spontaneous emission of $e^+ e^-$ pairs. In the paper [5.30], Schwinger calculated the critical field strengths E_S:

$$E_S = \frac{m_e^2 c^3}{e\hbar} . \tag{5.86}$$

Considering formula (5.86), we define the *Schwinger* force:

$$F_{\text{Schwinger}}^e = eE_S = \frac{m_e^2 c^3}{\hbar} . \tag{5.87}$$

Formula (5.87) can be written as:

$$F_{\text{Schwinger}}^e = \frac{m_e c}{\tau_{\text{Sch}}} , \tag{5.88}$$

where

Table 5.2. *Schwinger, Planck, Yukawa,* and *Bohr* forces

$F_{Schwinger}^e$	F_{Planck}	F_{Yukawa}	F_{Bohr}
(GeV/fm)	(GeV/fm)	(GeV/fm)	(GeV/fm)
$\sim 10^{-6}$	$\sim 10^{38}$	$\sim 10^{-2}$	$\sim 10^{-13}$

$$\tau_{\text{Sch}} = \frac{\hbar}{m_e c^2} \tag{5.89}$$

is *Schwinger* relaxation time for the creation of $e^+ e^-$ pair. Considering formulae (5.75), the relation of F_{Yukawa} and F_{Bohr} to the *Schwinger* force can be established

$$F_{\text{Yukawa}} = \alpha_s^3 \left(\frac{m_N}{m_e}\right)^2 F_{\text{Schwinger}}^e, \qquad \alpha_s = 0.15, \tag{5.90}$$

$$F_{\text{Bohr}} = \alpha_{em}^3 F_{\text{Schwinger}}^e, \qquad \alpha_{em} = \frac{1}{137},$$

and for *Planck* force

$$F_{\text{Planck}} = \left(\frac{M_{\text{P}}}{m_e}\right)^2 F_{\text{Schwinger}}^e. \tag{5.91}$$

In Table 5.2, the values of the $F_{\text{Schwinger}}^e$, F_{Planck}, F_{Yukawa}, and F_{Bohr} are presented, all in the same units GeV/fm. As in those units the forces span the range 10^{-13} to 10^{38}, it is valuable to recalculate the *Yukawa* and *Bohr* forces in the units natural to nuclear and atomic levels. In that case, one obtains:

$$F_{\text{Yukawa}} \sim 16 \text{ MeV/fm} . \tag{5.92}$$

It is quite interesting that $a_v \sim 16$ MeV is the volume part of the binding energy of the nuclei (droplet model).

For the *Bohr* force, considering formula (5.75), one obtains:

$$F_{\text{Bohr}} \sim \frac{50 \text{ eV}}{0.1 \text{ nm}} . \tag{5.93}$$

Considering that the Rydberg energy ~ 27 eV and *Bohr* radius ~ 0.1 nm formula (5.93), can be written as

$$F_{\text{Bohr}} \sim \frac{\text{Rydberg energy}}{\text{Bohr radius}} . \tag{5.94}$$

References

5.1.   J. Marciak-Kozlowska, M. Kozlowski: Found. Phys. Lett. **9**, 235 (1996)

5.2.   D. Jou, J. Casas-Vásquez, G. Lebon: *Extended Irreversible Thermodynamics* (Springer-Verlag, Berlin, 1993)

5.3.   R. Maartens: Class. Quantum Grav. **12**, 1455 (1995)

5.4.   P.G. Bergman: *Cosmology and Particle Physics,* ed by V. de Sabbata, H. Tso-Hain (Kluwer Academic Dordrecht 1994) p 9

5.5.   M. Kozlowski, J. Marciak-Kozlowska: Found. Phys. Lett. **10**, 295 (1999)

5.6.   E. Nelson: Phys. Rev. **150**, 1079 (1966)

5.7.   D.D. Joseph: *Fluid Dynamics of Viscoelastic Liquids* (Springer-Verlag New York 1990) p 71

5.8.   A.G. Riess et al.: Astron. J. **116**, 1009 (1998)

5.9.   G. Starkman et al.: Phys. Rev. Lett. **83**, 1510 (1999)

5.10. M. Kozlowski, J. Marciak-Kozlowska: Found. Phys. Lett. **10**, 599 (1997)

5.11. P.A.M. Dirac: Nature (London) **139**, 323 (1937)

5.12. T. Damour, G.W. Gibbons, C. Gundach: Phys. Rev. Lett. **64**, 123 (1990)

5.13. T. Damour, G. Esposito Farèse: Classical Quantum Gravity **9**, 2093 (1992)

5.14. La D., P.J. Steinhard: Phys. Rev. Lett. **62**, 376 (1989)

5.15. M. Gasperini: Gen. Rel. Grav. **30**, 1703 (1998)

5.16. G. Barton: in *Elements of Green's Functions and Propagation* (Oxford Science Publications Clarendon Press Oxford 1995) p 222

5.17. M. Maimonides: (1190), in *The Guide for the Perplexed* (transl. M. Friedlander George Routledge London 1904) p 121

5.18. R. Descartes: Meditions on the first philosophy. In *A Discourse on Method etc.* (transl. A.D. Lindsay Deut London 1912) pp 107–108

5.19. G.J. Whitrow: In *The Natural Philosophy of Time*, 2nd ed, (Oxford Science Publications 1990) p 204

5.20. A.H. Guth: In *The Inflationary Universe: The Guest for a New Theory of Cosmic Origins* (Addison-Wesley New York 1977); A.H. Guth: PNAS **90**, 4871 (1993)

5.21. R. Cayrel et al.: Nature **409**, 691 (2001)

5.22. D.N. Spergel et al.: PNAS **94**, 6579 (1997)

5.23. J.D. Anderson et al.: Phys. Rev. Lett. **81**, 2858 (1998)

5.24. F. Calogero: Phys. Lett. **A228**, 335 (1997)

5.25. M. Kozlowski, J. Marciak-Kozlowska: Nuovo Cimento **116B**, 821 (2001)

5.26. J.D. Anderson et al.: Phys. Rev. Lett. **81**, 2858 (1998)

5.27. M. Planck: *The Theory of Heat Radiation*, (Dover Publications 1959) p 173

5.28. H. Yukawa H, Proc. Phys.-Math. Soc. Japan **17**, 48 (1935)

5.29. TESLA Technical Design Report, http://tesla.desy.de/new-pages/TDR-CD/start.html

5.30. J. Schwinger: Phys. Rev. **82**, 664 (1951)

6

Attophysics and Technology with Ultrashort Laser Pulses

6.1 Quantum Heat Transport: From Basics to Applications

6.1.1 Ranges of Interactions and Heaton Energies for Quark, Electron, and Nucleon Gases

The development of ultraintense laser pulses will allow the study of new regimes of laser-matter interaction [6.1]. Lasers are now being designated [6.2] that will eventually lead to light intensities, such that $I\lambda_\mu^2 \gg 10^{19}\,\mathrm{W\mu m^2/cm^2}$. Here I is the laser intensity of the laser light and λ_μ is the wavelength in microns. In such intensities, the electron jitter velocity in the laser electric field becomes relativistic: $p_0/mc > 1$, where p_0 is jitter momentum, m is electron rest mass, and c is the light velocity in vacuum. When such lasers interact with an overdense plasma, it has been shown that a large number of relativistic superthermal electrons with energy E_{hot}

$$E_{\mathrm{hot}} \sim \left[\sqrt{1 + \frac{I\lambda_\mu^2}{1.4\,10^{18}}} - 1 \right] mc^2 \tag{6.1}$$

are produced [6.3]. Hence, $E_{\mathrm{hot}} > mc^2$ for $I\lambda_\mu^2 > 4\,10^{18}$. For even higher $I\lambda_\mu^2$, E_{hot} can exceed the pair $(\mathrm{e^+, e^-})$ production threshold. In the result, the interaction of ultraintense laser beams with matter can produce copious electron-positron pair, which represents a new state of matter with new thermal and radiative properties drastically different from ordinary plasma [6.4]. For the moment, the production of electron-positron pair was realized in the SLAC experiment [6.5]. In that experiment, a signal of 106±14 positrons above background has been observed in collisions of a low-emittance 46.6 GeV electron beam with terawatt pulse from a Nd: glass laser at 527 nm wavelength.

The positrons are interpreted as arising from a two-step process in which laser photons are backscattered to GeV energies by the electron beam followed by a collision between the high-energy photon and several laser photons to produce an electron-positron pair.

The creation of superthermal electron-positron pair is a relativistic effect, both because of the conversion of mass $\Leftrightarrow$ energy and the relativistic energies of created particle-antiparticle pairs. The natural frame to analyze the relativistic gases of particles is the quantum heat transfer equation (QHT) [6.6].

The GeV energy of laser photons emitted in the SLAC experiment are precursors of a new field of interdisciplinary applications of laser femtosecond beams. This superenergetic photons can, in principle, create not only electron and nucleon fermionic gases but also the free quark-gluon gas (if it exists!).

In Chapter 3, the quantum heat transport equation (QHT) was formulated. For electron and nucleon gases, the QHT has the form:

for electrons:

$$\tau^e \frac{\partial^2 T^e}{\partial t^2} + \frac{\partial T^e}{\partial t} = \frac{\hbar}{m_e} \nabla^2 T^e, \tag{6.2}$$

for nucleons:

$$\tau^N \frac{\partial^2 T^N}{\partial t^2} + \frac{\partial T^N}{\partial t} = \frac{\hbar}{m} \nabla^2 T^N. \tag{6.3}$$

In equations (6.2) and (6.3), m_e and m are the masses of electron, and nucleon respectively, and

$$\tau^e = \frac{\hbar}{m_e \alpha^2 c^2}, \qquad \tau^N = \frac{\hbar}{m(\alpha^s)^2 c^2}, \tag{6.4}$$

where τ^e and τ^N are the relaxation times for electrons and nucleons, respectively. The constants $\alpha = e^2/\hbar c$, $\alpha^s = m_\pi/m$ (m_π denotes the π meson mass), are the fine-structure constants for electromagnetic and strong interactions.

As was shown in Chapter 3, static spherically symmetric solutions of (6.2) and (6.3) potentials have the form

$$V^e(r) = -\frac{g^e}{r} e^{-\frac{r}{R_e}}, \qquad g^e = \alpha \hbar c, \tag{6.5}$$

$$V^N(r) = -\frac{g^N}{r} e^{-\frac{r}{R_N}}, \qquad g^N = \alpha^s \hbar c, \tag{6.6}$$

where the ranges of electromagnetic interaction (in solids) and strong interaction in nucleus equal

$$R_e = \frac{2\hbar}{m_e \alpha c}, \tag{6.7}$$

$$R_N = \frac{2\hbar}{m \alpha^s c}. \tag{6.8}$$

The potentials $V^e(r)$, $V^N(r)$ are the Debye-Hückl potential and Yukawa potential, respectively [6.6].

It is quite natural to pursue the study of the thermal excitation to the subnucleon level, i.e. quark matter. Analogously as for electron and nucleon gases, for quark gas the QHT has the form

$$\frac{1}{c^2}\frac{\partial^2 T^q}{\partial t^2} + \frac{1}{c^2\tau}\frac{\partial T^q}{\partial t} = \frac{(\alpha_s^q)^2}{3}\nabla^2 T^q, \tag{6.9}$$

with α_s^q the fine-structure constant for strong quark-quark interaction. In paper [6.7], α_s^q was calculated, $\alpha_s^q = 1$. The *heaton* energy [6.6] for quark gas can be defined as

$$e_h^q = \frac{m_q}{3}(\alpha_s^q)^2 c^2, \tag{6.10}$$

where m_q denotes the average quark mass and $m_q = 417$ MeV [6.7]. With formula (6.10), the *heaton* energy for quark gas equals

$$e_h^q \cong 139\,\text{MeV} = m_\pi. \tag{6.11}$$

It occurs that when we attempt to "melt" the nucleons in order to obtain the free quark gas, the energy of the *heaton* is equal to the π-meson mass. This is the thermodynamics presentation for the quark *confinement*. Moreover, we conclude that only from hyperbolic quantum heat transport equation we obtain finite mass for particle that mediates the strong interaction. For parabolic heat transport equation $\tau^N \to 0$, i.e. (6.4)

$$\tau^N = \frac{\hbar}{m(\alpha^s)^2 c^2} = \frac{\hbar m}{m_\pi^2 c^2} \to 0, \tag{6.12}$$

hence, $m_\pi \to \infty$. Analogously from Fourier equation one can conclude that due to the fact that $v_h \to \infty$, all interactions must have zero range as from (6.7) and (6.8)

$$R^{e,N,q} = \frac{2\hbar}{mv_h^{e,N,q}} \to 0, \quad \text{when} \quad v_h \to \infty.$$

In Table 6.1, the results for calculations of the ranges of interactions and *heaton* energies for electron, nucleon and quark gases are presented. From the inspection of Table 6.1, we conclude that the Fourier equation cannot be applied to study of the thermal processes on the atomic, nuclear, and quark scales.

Table 6.1. Ranges of Interactions and *Heaton* Energies for Quark, Electron and Nucleon Gases

	Range (m)		*Heaton* Energy (eV)	
Particles	QHT	Fourier	QHT	Fourier
Quarks	10^{-16}	0	$1.39\,10^8$	∞
Electrons	10^{-10}	0	9	∞
nucleons	10^{-15}	0	$7\,10^6$	∞

6.1.2 Hierarchical Structure of the Thermal Excitation

Recently, the new field of thermal investigations of nanoparticles (i.e. particles with radius r of the order nanometer) was developed [6.9]. In paper [6.10], the model for the relaxation of thermal excitation of the nanoparticles was derived and obtained to the study of Ga nanoparticles. It was shown that the thermodynamical properties of nanoparticle depend on its geometrical dimensions.

The fact that the femtosecond thermodynamical properties of nanoparticles depend on fine-structure constant α opens the question: is the constant α is really constant? Recently [6.11], the search for fine-structure constant was undertaken with positive results. It occurs that $\Delta\alpha/\alpha \sim 10^{-5}$.

In the paper [6.10], starting with atomic values of the relaxation time τ_e and velocity of thermal wave v_h, the microscopic model of the relaxation processes in nanoparticles was formulated. Considering the quantum heat transport equation

$$\tau^e \frac{\partial^2 T^e}{\partial t^2} + \frac{\partial T^e}{\partial t} = \frac{\hbar}{m_e}\nabla^2 T^e \tag{6.13}$$

and Pauli-Heisenberg inequality

$$\Delta r\,\Delta p \geq N^{1/3}\hbar \tag{6.14}$$

the thermal velocity v_h^f and relaxation time τ^f were calculated for nanoparticles:

$$v_h^f = \frac{1}{N^{1/3}}v_h\,, \tag{6.15}$$

$$\tau^f = N\tau\,. \tag{6.16}$$

In formulas (6.15) and (6.16), N denotes the number of particles ("partons")
in a nanoparticle. For a nanoparticle with radius r, the number of "partons"
equals

$$N = \frac{\frac{4\pi}{3} r^3 \varrho A Z}{\mu},\tag{6.17}$$

where ϱ is the density of nanoparticles, A is the Avogadro number, Z is the
number of the valence electrons, and μ is the molecular mass of the nanoparticle material.

Two new interesting results can be concluded from (6.15)–(6.17). First of
all, the radius of a nanoparticle is proportional to the number of *partons* [6.9]:

$$r \sim N^{1/3}.\tag{6.18}$$

In this aspect, the nanoparticle resembles the atomic nucleus in which [6.2]:

$$r \sim A^{1/3}.\tag{6.19}$$

In that case, A is the mass number of the nucleus (but not Avogadro number!).
The formula (6.16) describes the quantization rule for the relaxation time. It
can be stated that the τ (atomic relaxation time) is the *quantum* of relaxation
time.

From the theoretical point of view there emerges a model of "free" electrons
in clusters as a system of Fermi (spin one-half) particles being quantized
in a global mean field. In a nanoparticle, the global mean field is not the
screened Coulomb potential around the positive charge of the point as it is in
the nucleus; it is more like cavity [6.10]. And the positive charge is smeared
out through the whole volume. The nanoparticles constitute a new family of
what may be called *quasi-atoms* or even more descriptively *giant atoms*. The
analogy between real atoms and metallic nanoparticles has its limits. First,
the clusters can easily lose single atoms or be split into smaller clusters. They
are not indivisible the way atoms are. Unlike the atoms, where the positive
charges are extremely hard frozen, the ionic charges in a cluster may be highly
excited thermally. As a result, the quantized electronic motion is really taking
place in a heat bath. The electrons may be thermally excited (for example
by ultrashort laser pulses) out of their ground state [6.10] configuration and
eventually build the thermal waves penetrating the nanoparticles.

At closer look, the giant atoms are conceptually hybrids between atomic
and nuclear quantum systems. The constant density (ϱ) and the deformability

Table 6.2. Hierarchical Structure of the Thermal Excitation

Hierarchical Structures	τ (s)	v_h (m/s)	E_h (eV)	References
Atom	$\dfrac{\hbar}{m_e v_h^2}$	$\alpha c,\ \alpha = \dfrac{e^2}{\hbar c}$	$m_e v_h^2$	(a)
Molecule	$\dfrac{m_p}{m_e}\dfrac{\hbar}{m_e v_h^2}$	$\alpha c(\dfrac{m_e}{m_p})^2$	$\alpha^2 \dfrac{m_e}{m_p} m_e c^2$	(b)
Nanoparticle containing N particles	$N\dfrac{\hbar}{m_e v_h^2}$	$\dfrac{1}{N^{1/3}}\alpha c$	$\dfrac{m_e}{N^{2/3}}\alpha^2 c^2$	(c)
Atomic nucleus	$\dfrac{\hbar}{m_N v_h^2}$	$\alpha^s c$	$m_p(\alpha^S c)^2$	(d)
		$\alpha^s = \dfrac{2m_e}{\alpha m_p}$		

(a) J. Marciak-Kozlowska, M. Kozlowski: Lasers Eng. **5**, 79 (1996);
(b) M. Kozlowski, J. Marciak-Kozlowska: Lasers Eng. **9**, 103 (1999);
(c) M. Kozlowski, J. Marciak-Kozlowska: Lasers Eng. **10**, 37 (2000);
(d) M. Kozlowski, J. Marciak-Kozlowska: Lasers Eng. **7**, 13 (1998).

are nuclear characteristic but the quantized constituents are electrons, as in atom. The *quasi-atom* can be looked at as an ordinary atom differing only by having the nuclear charge distributed essentially through the whole atom.

In Table 6.2, the hierarchical structure of the matter is presented. It occurs that starting with nucleus, through the atom, the molecule, and the nanoparticle, relaxation time, velocity of heat propagation, and *heaton* energy can be described with the help of two constants $\alpha = e^2/\hbar c = 1/137$ and $\beta = m_e/m_p = 1/1836$.

We conclude that despite the vast complexity of everything made out of atoms and molecules, the thermal properties of all entities are determined by the values of just two numbers. Up to now we do not know why these two numbers take the precise values that they do. Were they different, our Universe would be different, perhaps unimaginably different [6.12]. In paper [6.13], the "ambient temperature" important for biological cells was calculated

$$T_m \sim 10^{-3}\alpha^2 m_e c^2 \sim 316 \text{ K} = 40^0 \text{ C}. \qquad (6.20)$$

In Fig. 6.1(a), the values of T_m for different α are presented. When α changes in the range -10% to 10%, the ambient temperature changes from 20^0C to 100^0C. The problem "How constants are constants" is not an academic one. A time-varying fine structure constant, α, can now be sought with

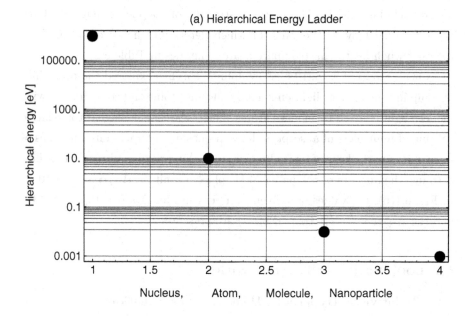

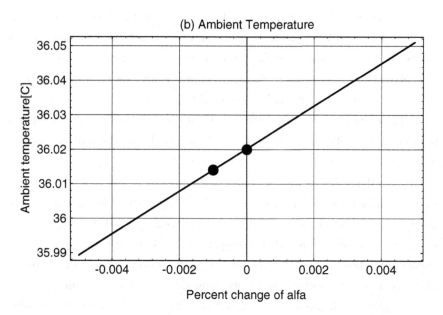

Fig. 6.1. (a) Hierarchy of thermal excitation in nucleus, atom, molecule, and nanoparticle. (b) The ambient temperature as the function of the change of fine structure constant. The experimental results of paper [6.11] is also presented (lower circle).

a new technique [6.11]. The inherent strength of the electromagnetic force is characterized by α, the value of which determines how well atoms hold together when heated up (compare heaton energies in Table 6.2).

Now, a group of scientists led by John Webb has explored the possibility of sampling the ancient light emitted by ancient atoms and comparing it with modern light emitted by modern atoms. The researchers looked at the relative spacing of multiplets of absorption lines in quasar spectra, comparing lines of ionized iron with those of ionized magnesium. After taking all corrections into account, the authors of the paper [6.11] place a limit on $\Delta\alpha/\alpha \sim -1.1\ 10^{-5}$. In Fig. 6.1(b), the values of ambient temperature for $\Delta\alpha/\alpha = 0$ ($\alpha = 1/137$) and the $\Delta\alpha/\alpha = 10^{-3}\%$ (paper [6.11]) are presented.

6.2 Looking into the Nanoworld

6.2.1 Relativistic Hyperbolic Heat Transfer Equation

With the advent of attosecond lasers (1 as = 10^{-18} s), physicist, and engineers are close to controlling the motion of electrons on a time scale that is substantially shorter than the oscillation period of visible light. It is also possible with attosecond laser pulse to rip an electron wave-packet from the core of an atom and set it free with similar temporal precision.

Because we know that there is an absolute velocity of light, it is sufficient to define the smallest time interval (e.g. 1 as), and thus the unit of length is a derivative quantity (3 10^{-10} m = 0.3 nm = Bohr radius).

After a definition of time and space, the laws of classical physics relating such parameters as distance, time, velocity, temperature are assumed to be independent of accuracy with which these parameters can be measured. It should be pointed out that this assumption is not explicitly stated in the formulation of classical physics. This implies that together with the assumption of the existence of the objective reality independently of any measurements (in classical physics), it was tacitly assumed that *there is a possibility of unlimited increase in accuracy of measurements.* Bearing in mind the atomicity of time, i.e. considering the smallest time period, Planck time, the above statement is obviously not true. With attosecond laser pulses, we are at the limit of laser time resolution.

With the advent of attosecond laser pulses, we enter into a new Nanoworld where size becomes comparable to atomic dimension, where transport phenomena follow different laws from those in a macro world. This signifies the end of first stage of miniaturization (from 10^{-3} m to 10^{-6} m) and the beginning of a new one (from 10^{-6} to 10^{-9}). Nanoworld is a quantum world with all predictable and still unpredictable features.

In the subsequent, we develop and solve the quantum relativistic heat transport equation for Nanoworld transport phenomena in the presence of external forces. This is the generalization of the results of paper [6.14] in which quantum relativistic hyperbolic equation was proposed and solved.

In paper [6.14], relativistic hyperbolic transport equation (RHT) was formulated:

$$\frac{1}{v^2}\frac{\partial^2 T}{\partial t^2} + \frac{m_0\gamma}{\hbar}\frac{\partial T}{\partial t} = \nabla^2 T. \tag{6.21}$$

In equation (6.21), v is the velocity of heat waves, m_0 is the mass of heat carrier, and γ – the Lorentz factor $\gamma = (1 - \frac{v^2}{c^2})^{-1/2}$. As was shown in paper [6.14], the heat energy (heaton temperature) T_h can be defined as follows:

$$T_h = m_0\gamma v^2. \tag{6.22}$$

Considering that v, thermal wave velocity equals [6.14]

$$v = \alpha c, \tag{6.23}$$

where α is the coupling constant for the interactions that generate the thermal wave ($\alpha = 1/137$ and $\alpha = 0.15$ for electromagnetic and strong force, respectively), heaton temperature equals

$$T_h = \frac{m_0\alpha^2 c^2}{\sqrt{1-\alpha^2}}. \tag{6.24}$$

From formula (6.24), one concludes that heaton temperature is the linear function of the mass m_0 of the heat carrier. It is quite interesting to observe that the proportionality of T_h and heat carrier mass m_0 was the first time observed in ultrahigh energy heavy-ion reactions measured at CERN [6.15]. In paper [6.15], it was shown that temperature of pions, kaons, and protons produced in Pb+Pb, S+S reactions are proportional to the mass of particles. Recently at Rutherford Appleton Laboratory (RAL), the VULCAN laser was used to produce the elementary particles: electrons and pions [6.16].

In paper [6.17], the damped thermal wave equation was developed:

$$\frac{1}{v^2}\frac{\partial^2 T}{\partial t^2} + \frac{m}{\hbar}\frac{\partial T}{\partial t} + \frac{2Vm}{\hbar^2}T - \nabla^2 T = 0. \tag{6.25}$$

The relativistic generalization of (6.25) is quite obvious:

$$\frac{1}{v^2}\frac{\partial^2 T}{\partial t^2} + \frac{m_0\gamma}{\hbar}\frac{\partial T}{\partial t} + \frac{2V m_0\gamma}{\hbar^2}T - \nabla^2 T = 0\,. \tag{6.26}$$

It is worthwhile to note that in order to obtain nonrelativistic equation, we put $\gamma = 1$.

If the nucleus contains Z protons, then we have for electron temperature in atom

$$T_h^Z(\text{atom}) = m_e(Z\alpha)^2 c^2\,. \tag{6.27}$$

Equation (6.27) also tells us the magnitude of the temperature of the ionization of the atom:

$$T_h^Z(\text{ionization}) = \frac{1}{2}m_e c^2\,(Z\alpha)^2\left(1 + \frac{m_e}{m_p}\right)^{-1} \sim Z^2 10^5\text{K}\,.$$

The motion of electron in the atom is equivalent to the flow of an electric current in a loop of wire. With attosecond laser pulses, we will be able to influence the current in the atomic "wire." This opens quite new perspective for the Nanoworld technology, or nanotechnology. The new equation (6.26) is the natural candidate for the master equation that can be used in nanotechnology.

When external force is present, $F(x,t)$ the forced damped heat transport is obtained instead of (6.26) (in one-dimensional case):

$$\frac{1}{v^2}\frac{\partial^2 T}{\partial t^2} + \frac{m_0\gamma}{\hbar}\frac{\partial T}{\partial t} + \frac{2V m_0\gamma}{\hbar^2}T - \frac{\partial^2 T}{\partial x^2} = F(x,t)\,. \tag{6.28}$$

The hyperbolic relativistic quantum heat transport equation (RQHT), (6.28), describes the forced motion of heat carriers that undergo the scatterings ($\frac{m_0\gamma}{\hbar}\frac{\partial T}{\partial t}$ term) and are influenced by potential ($\frac{2V m_0\gamma}{\hbar^2}T$ term).

The solution of equation can be written as

$$T(x,t) = e^{-t/2\tau}u(x,t)\,, \tag{6.29}$$

where $\tau = \hbar/(mv^2)$ is the relaxation time. After substituting formula (6.29) into (6.28), we obtain new equation

$$\frac{1}{v^2}\frac{\partial^2 u}{\partial t^2} - \frac{\partial^2 u}{\partial x^2} + qu(x,t) = e^{\frac{t}{2\tau}}F(x,t)\,, \tag{6.30}$$

and

$$q = \frac{2Vm}{\hbar^2} - \left(\frac{mv}{2\hbar}\right)^2\,, \tag{6.31}$$

$$m = m_0\gamma\,.$$

Equation (6.30) can be written as:

$$\frac{\partial^2 u}{\partial t^2} - v^2 \frac{\partial^2 u}{\partial x^2} + qv^2 u(x,t) = G(x,t),$$ (6.32)

where

$$G(x,t) = v^2 e^{\frac{t}{2\tau}} F(x,t).$$

When $q > 0$, (6.32) is the forced Klein-Gordon equation K-GE. The solution of the forced Klein-Gordon equation for the initial conditions:

$$u(x,0) = f(x), \qquad u_t(x,0) = g(x)$$ (6.33)

has the form [6.17]:

$$u(x,t) = \frac{f(x-vt) + f(x+vt)}{2}$$ (6.34)

$$+ \frac{1}{2v} \int_{x-vt}^{x+vt} g(\zeta) J_0 \left[q\sqrt{v^2 t^2 - (x-\zeta)^2} \right] d\zeta$$

$$- \frac{\sqrt{q}\, vt}{2} \int_{x-vt}^{x+vt} f(\zeta) \frac{J_1 \left[q\sqrt{v^2 t^2 - (x-\zeta)^2} \right]}{\sqrt{v^2 t^2 - (x-\zeta)^2}} d\zeta$$

$$+ \frac{1}{2v} \int_0^t \int_{x-v(t-t')}^{x+v(t-t')} G(\zeta,t') J_0 \left[q\sqrt{v^2 (t-t')^2 - (x-\zeta)^2} \right] dt' d\zeta.$$

When $q < 0$, equation (6.32) is the forced modified Heaviside (telegraph) equation with the solution: [6.17]

$$u(x,t) = \frac{f(x-vt) + f(x+vt)}{2}$$ (6.35)

$$+ \frac{1}{2v} \int_{x-vt}^{x+vt} g(\zeta) I_0 \left[-q\sqrt{v^2 t^2 - (x-\zeta)^2} \right] d\zeta$$

$$+ \frac{v\sqrt{-q}\, t}{2} \int_{x-vt}^{x+vt} f(\zeta) \frac{I_1 \left[-q\sqrt{v^2 t^2 - (x-\zeta)^2} \right]}{\sqrt{v^2 t^2 - (x-\zeta)^2}} d\zeta$$

$$+ \frac{1}{2v} \int_0^{t'} \int_{x-v(t-t')}^{x+v(t-t')} G(\zeta,t') I_0 \left[-q\sqrt{v^2 (t-t')^2 - (x-\zeta)^2} \right] d\zeta dt'.$$

When $q = 0$ (6.32) is the forced thermal equation [6.17].

$$\frac{\partial^2 u}{\partial t^2} - v^2 \frac{\partial^2 u}{\partial x^2} = G(x,t). \tag{6.36}$$

On the other hand one can say that (6.36) is the distortionless hyperbolic equation. The condition $q = 0$ can be rewritten, as:

$$V\tau = \frac{\hbar}{8}. \tag{6.37}$$

The equation (6.37) is analogous to the Heisenberg uncertainty relations. Considering (6.22), equation (6.37) can be written as:

$$V = \frac{T_h}{8}, \qquad V < T_h. \tag{6.38}$$

One can say that the distortionless waves can be generated only if $T_h > V$. For $T_h < V$, i.e. when the "Heisenberg rule" is broken, the shape of the thermal waves is changed.

We consider the initial and boundary value problem for the inhomogenous thermal wave equation in semi-infinite interval [6.17]: that is

$$\frac{\partial^2 u}{\partial t^2} - v^2 \frac{\partial^2 u}{\partial x^2} = G(x,t), \qquad 0 < x < \infty, \qquad t > 0, \tag{6.39}$$

with initial condition

$$u(x,0) = f(x), \qquad \frac{\partial u(x,0)}{\partial t} = g(x), \qquad 0 < x < \infty,$$

and boundary condition

$$au(0,t) - b\frac{\partial u(0,t)}{\partial x} = B(t), \qquad t > 0, \tag{6.40}$$

where $a \geq 0$, $b \geq 0$, $a + b > 0$ (with a and b both equal to constants) and F, f, g, and B are given functions. The solution of (6.39) is of the form [6.17]:

$$u(x,t) = \frac{1}{2}\left[f(x - vt) + f(x + vt)\right] + \frac{1}{2v}\int_{x-vt}^{x+vt} g(s)ds \tag{6.41}$$

$$+ \frac{1}{2v}\int_0^t \int_{x-v(t-t')}^{x+v(t-t')} F(s,t')dsdt'.$$

In the special case where $f = g = F = 0$, we obtain the following solution of the initial and boundary value problem (6.39, 6.40):

$$u(x,t) = \begin{cases} 0, & x > vt, \\ \dfrac{v}{b} \displaystyle\int_0^{t-\frac{x}{v}} \exp\left[\dfrac{va}{b}\left(y - t + \dfrac{x}{v}\right)\right] B(t)dt, & 0 < x < vt, \end{cases} \tag{6.42}$$

if $b \neq 0$. If $a = 0$ and $b = 1$, we have:

$$u(x,t) = \begin{cases} 0, & x > vt, \\ v \displaystyle\int_0^{t-\frac{x}{v}} B(y)dy, & 0 < x < vt, \end{cases}$$

whereas if $b = 0$ and $a = 1$, we obtain:

$$u(x,t) = \begin{cases} 0, & x > vt, \\ B\left(t - \dfrac{x}{v}\right), & 0 < x < vt. \end{cases}$$

It can be concluded that the boundary condition (6.40) gives rise to a wave of the form $K(t - \frac{x}{v})$ that travels to the right with speed v. For this reason, the foregoing problem is often referred to as a *signaling problem* for the thermal waves.

With attosecond laser pulses, we expect to be able to track the motion of an electron within a femtosecond. We could also follow the relaxation of the remaining bound electrons to their new equilibrium states. Once we can generate attosecond pulses with much higher energies, we will be able to use them as both pumps and probes in pump-probe spectroscopy on the atomic level.

Up to now, the pump-probe method allows the investigation of the heat transport on the bulk scale. The experiments triggered by the Brorson paper [6.19] disclosed the finite velocity of the heat transport, i.e. the hyperbolicity of heat transport equation.

Accordingly, with the attosecond pump-probe method, we will be able to study heat transport in the Nanoworld. As was shown in [6.14], one can define the electron temperature (heaton temperature) in ground state hydrogen-like atom:

$$T_h(\text{atom}) = m_e \alpha^2 c^2, \tag{6.43}$$

and in hydrogen-like molecule [6.13, 6.20]

$$T_h(\text{molecule}) = \alpha^2 \frac{m_e}{m_p} m_e c^2. \tag{6.44}$$

6.2.2 Laser Melting of Nanoparticles with Negative Heat Capacity

A system that gets colder when energy is added is said to have a negative heat capacity and it can attain thermal equilibrium with its surroundings. Normally, energy flows into cold objects and their temperature rises. If the system has a negative heat capacity, however, the object will get progressively colder.

Recently, in a paper by M. Schmidt et al. [6.21], it was shown that a nanoparticle of 147 Na^+ atoms heated with a laser beam has a negative heat capacity.

The main results can be summarized as follows [6.21]:

1. Sodium clusters $^{147}Na^+$ can be thermalized, which gives them a canonical distribution of internal energies. One cluster size is selected.
2. The selected cluster is irradiated with laser and the distribution of photofragments is measured as a function of the cluster temperature. From this, the total internal energy E_t of the cluster can be determined.
3. The microcanonical heat capacity is negative near T_{molt}.

In this section the model for the heating process of nanoparticles will be proposed. Assuming that the ionized atoms and electrons in a nanoparticle attract each other by Coulomb type forces, the velocity and temperature of the ion in the nanoparticle can be calculated. It will be shown that for systems with an attractive potential of the type $V(r) = \alpha/r$ where α is the strength of the attractive force and when the temperature field is quantized, $T \propto \alpha^2$ (*heaton* temperature [6.22]), the heat capacity is negative, $C_v < 0$.

The introduction of long-range interaction (Coulomb interaction) into the thermodynamics of nanoparticles may bring about unexpected results. The interaction of ions and electrons embedded in a nanoparticle can be described as

$$F_c = \frac{\alpha \hbar c}{r^2}, \qquad (6.45)$$

where α is the electromagnetic fine structure constant ($\alpha = 1/137$), $\hbar$ is the Planck constant, and c is the speed of light. For the central force (6.45), the atom velocity can be calculated from the formula

$$\frac{mv^2}{r} = \frac{\alpha \hbar c}{r^2} \qquad (6.46)$$

$$v = \frac{(\alpha \hbar c)^{1/2}}{(mr)^{1/2}}.$$

In formula (6.46), m is the mass of the ions embedded in the nanoparticle.

The total energy E_t of an ion in the nanoparticle is described by the equation

$$E_t = \frac{mv^2}{2} - \frac{\alpha \hbar c}{r}. \tag{6.47}$$

Considering formula (6.46), one obtains

$$E_t = -\frac{1}{2}\frac{\alpha \hbar c}{r}. \tag{6.48}$$

Note that the total energy is negative. Hence, a large r means greater energy, a small r, less energy.

If we give the nanoparticle some energy (through heating by laser beam) then r must become larger – ions can evaporate. Hence if the nanoparticle is heated, then the velocity will become smaller. Assuming as a first approximation the classical gas of ions:

$$\frac{mv^2}{2} = \frac{3}{2}k_BT, \tag{6.49}$$

where k_B denotes the Boltzmann constant, $k_B = 8.67 \cdot 10^5$ eVk^{-1}, one obtains:

$$T = \frac{1}{3}mv^2 = \frac{1}{3}\frac{\alpha \hbar c}{r k_B}. \tag{6.50}$$

The formula describes the *heaton* [6.22] energy for the gas of ions in nanoparticle.

Substituting (6.50) into formula (6.48), we obtain:

$$E_t = -\frac{3}{2}k_BT. \tag{6.51}$$

We define heat capacity (per atom) of a nanoparticle as

$$C_V = \frac{\partial E_T}{\partial t} \tag{6.52}$$

and from formula (6.51) one obtains:

$$C_V = -\frac{3}{2}k_B. \tag{6.53}$$

Formula (6.53) describes the new result concerning the thermodynamic properties of nanoparticles at the melting temperature. The heat capacity is negative. It means that a nanoparticle is such a system that adding heat lowers the temperature and extracting heat raises the temperature. When sodium cluster heated by laser beam suddenly melts, its latent heat of fusion is removed from the kinetic energy of the system causing its temperature to drop. The overall effect is that the cluster has a negative heat capacity at its melting transition.

The existence of systems with a negative thermal capacity is catastrophic for thermodynamical properties. Consider the system shown in Fig. 6.2, which consists of part a with a positive C_V and part b with negative C_V.

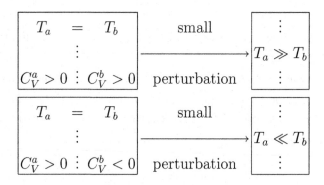

Fig. 6.2. For system with the negative heat capacity, thermal equilibrium is unstable.

In the beginning, the system is in thermal equilibrium and the temperatures $T_a = T_b$. The equilibrium is a dynamic one, that is, energy emitted by a is absorbed by b and *vice versa*. The two cancel out and equilibrium is maintained. There are always small fluctuations in an equilibrium. For example, the heat that a transfers to b may be slightly larger than what b gives to a, and b absorbs a little more energy. If b has a positive thermal capacity, then b's temperature will rise, its heat transfer to a will also increase and so cancel out the excess absorption of energy and return to equilibrium. However, if b has a negative thermal capacity, then an excess in the energy absorbed will lower the temperature. The b temperature will decrease continuously thus destroying the original equilibrium. On the contrary, where the fluctuation is such that b absorbs a little less energy, then the outcome is an ever-increasing temperature of b, again destroying the original equilibrium.

The conclusion is that for the system composed of a and b, thermal equilibrium is unstable and is destroyed by any slight fluctuation giving rise to temperature difference. Hence as long as self-attracting systems are present, a stable thermal equilibrium does not exist.

Negative heat capacities have been known in astrophysics [6.23], where energy can be added to a star or star cluster, which then cools down. A similar effect has been calculated [6.24] for fragmentating nuclei and has been observed as well [6.25].

Recently, H. Haberland and co-workers at the University of Freiburg have demonstrated the negativity of the heat capacity for nanoparticles [6.21]. They have shown that an isolated cluster (nanoparticle) of 147 sodium atoms has a negative heat capacity (at its melting point) [6.21]. How can this negative heat capacity be interpreted? Upon melting, a large system converts added energy completely into potential energy, reducing continuously the fraction of its solid phase. The kinetic energy and thus the temperature remain constant. Small cluster, on the other hand, tries to avoid partly molten states and prefers to convert some its kinetic energy into potential energy instead. Therefore, the cluster can become colder, while its total energy increases.

Melting in a macroscopic object occurs at some well-defined temperature. This is no longer true for a nanoparticle. Concepts like temperature, phase (solid, liquid) had originally been defined only for infinitely large systems. For the solid to liquid transition of finite systems (nanoparticles), one finds four main differences with respect to the bulk counterpart: (i) melting point is generally reduced, (ii) the latent heat is smaller, (iii) the transition does not occur at one definite temperature, and (iv) the heat capacity can become negative.

The main mechanism in melting for bulk and nanoscopic matter is diffusion. But as has been suggested [6.22], diffusion on a nanoscale is not Fourier diffusion. The diameter of a nanoparticle is of the order the atom mean free path and de Broglie length. In that case, the transport of thermal energy is described by the hyperbolic (not parabolic as in the Fourier equation) quantum heat transport equation, QHT [6.22]:

$$\tau\frac{\partial^2 T}{\partial t^2} + \frac{\partial T}{\partial t} = \frac{\hbar}{m}\nabla^2 T. \tag{6.54}$$

In equation (6.54), τ is the relaxation time [6.22]

$$\tau = \frac{\hbar}{m\alpha^2 c^2}, \tag{6.55}$$

where m is the ion mass, α is the strength of electromagnetic interaction, and c is light velocity. In [6.22], the solution of (6.54) was obtained and discussed. Equation (6.54) is then solved for the following initial boundary conditions:

$$\Delta T_0 = \frac{\beta \varrho_E}{C_V \, \alpha c \, \Delta t} \qquad \text{for} \qquad 0 \leq x < \alpha c \, \Delta t$$

$$\Delta T_0 = 0 \qquad \text{for} \qquad x > \alpha c \, \Delta t. \qquad (6.56)$$

Here ϱ_E denotes the heating pulse fluency, β is the efficiency of the absorption of energy in the nanoparticle, $C_V(T)$ is the nanoparticle heat capacity, and Δt the duration of the pulse. At $t = 0$, the temperature profile in the nanoparticle predicted by equation (6.56) is that of equation [6.22]

$$\Delta T(l,t) = \frac{1}{4} \Delta T_0 e^{-t/2\tau_a} \theta(t_0 + \Delta t - t) \qquad (6.57)$$

$$+ \frac{\Delta t}{4} \Delta T_0 e^{-t/2\tau_a} \left[I_0(z) + \frac{t}{2\tau} \frac{1}{z} I_1(z) \right] \theta(t - t_0),$$

where $z = (t^2 - t_0^2)^{1/2}/2\tau$ and $t_0 = l/\alpha c$.

In formula (6.57), $I_0(z)$, $I_1(z)$ are modified Bessel functions and $\theta(t - t_0)$ is the Heaviside function.

The solution of equation (6.54) when there are reflecting boundaries is the superposition of the temperature at l from the original temperature and from image heat source at $\pm 2nl$. The solution is

$$\Delta T(l,t) = \sum_{i=0}^{\infty} \Delta T_0 e^{-t/2\tau_a} \theta(t - t_i)\theta(t_i + \Delta t - t)$$

$$+ \Delta T_0 \frac{\Delta t}{2\tau} e^{-t/2\tau} \left[I_0(z_i) + \frac{t}{2\tau} \frac{1}{z_i} I(z_i) \right] \theta(t - t_i), \qquad (6.58)$$

where $t_i = t_0, 3t_0, 5t_0$.

The solution of equations (6.54) and (6.58) for $C_V > 0$ is presented in Fig. 6.3. Figure 6.3(a) describes the solution of hyperbolic heat transfer equation (6.54) and Fig. 6.3(b) the solution of the Fourier equation

$$\frac{\partial T}{\partial t} = D \frac{\partial^2 T}{\partial x^2}. \qquad (6.59)$$

The solution of equation (6.54) for $C_V < 0$ is presented in Fig. 6.4. Figure 6.4(a) describes the solution of hyperbolic heat transfer equation and Fig. 6.4(b) the solution of Fourier equation (6.59).

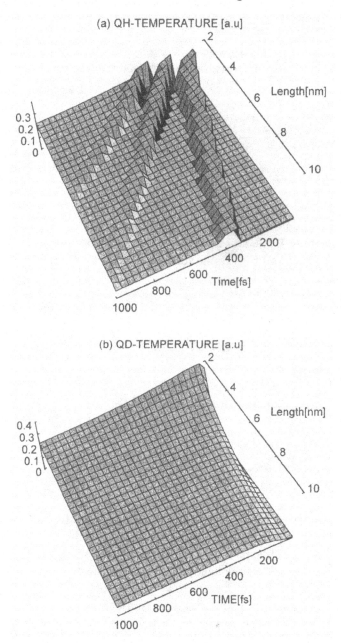

Fig. 6.3. The temperature field $T(x,t)$ for system with $C_V > 0$: (**a**) the solution of QHT, (**b**) the solution for PHT.

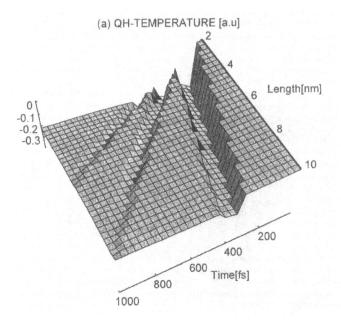

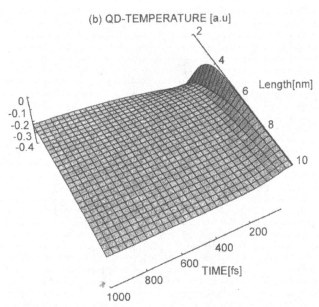

Fig. 6.4. The same as in Fig. 6.3 but for $C_V < 0$.

References

6.1. J.C. Diels, W. Rudolph: *Ultrashort Laser Pulse Phenomena* (Academic Press, Inc. San Diego 1996)

6.2. M.D. Perry, G. Mourou: Science **264**, 917 (1994)

6.3. E.P. Liang et al.: Phys. Rev. Lett. **81**, 4887 (1998)

6.4. R. Svenson et al.: Astrophys. J. **283**, 842 (1984)

6.5. D.L. Burke et al.: Phys. Rev. Lett. **79**, 1626 (1997)

6.6. M. Kozlowski, J. Marciak-Kozlowska: Lasers Eng. **7**, 81 (1998)

6.7. M. Kozlowski, J. Marciak-Kozlowska: Hadronic Journal **20**, 289 (1997)

6.8. B. Povh, K. Rith, Ch. Scholz, F. Zetsche: *Particles and Nuclei* (Springer-Verlag Berlin 1995)

6.9. M. Nisoli et al.: Phys. Rev. Lett. **78**, 3575 (1997)

6.10. M. Kozlowski, J. Marciak-Kozlowska: Lasers Eng. **10**, 37 (2000)

6.11. J.K. Webb et al.: Phys. Rev. Lett. **82**, 884 (1999)

6.12. J.D. Barrow: *Theories of Everything* (Fawcett Columbine New York 1991)

6.13. M. Kozlowski, J. Marciak-Kozlowska: Lasers Eng. **9**, 103 (1999)

6.14. J. Marciak-Kozlowska, M. Kozlowski: Lasers Eng. **11**, 259 (2001)

6.15. I.G. Bearden et al.: Phys. Rev. Lett. **78**, 2080 (1997)

6.16. K.W.D. Ledingham, P.A. Norreys: Contemporary Physics **40**, 367 (1999)

6.17. M. Kozlowski, J. Marciak-Kozlowska: Lasers Eng. **8**, 11 (1998)

6.18. E. Zauderer: *Partial Differential Equation of Applied Mathematics*, 2nd ed., (Wiley 1989)

6.19. S.D. Brorson et al.: Phys. Rev. Lett. **59**, 1962 (1987)

6.20. M. Kozlowski, J. Marciak-Kozlowska: *From Quarks to Bulk Matter* (Hadronic Press USA 2001)

6.21. M. Schmidt et al.: Phys. Rev. Lett. **86**, 1191 (2001)

6.22. M. Kozlowski, J. Marciak-Kozlowska: Lasers Eng. **5**, 79 (1996); **6**, 1 (1997); **7**, 13 (1998); **9**, 103 (1999); *From Quarks to Bulk Matter* (Hadronic Press USA 2001)

6.23. D. Lynden-Bell: Physica (Amsterdam) **263A**, 293 (1989)

6.24. D.H.E. Gross: Rep. Prog. Phys. Lett. **B53**, 65 (1990)

6.25. M. D'Agostino et al.: Phys. Lett. **B473**, 219 (2000)

6.26. G. Strobl: Eur. Phys. J. **E3**, 165 (2000)

6.27. K. Weslesen: Colloid. Polym. Sci. **278**, 608 (2000)

6.28. A.F. Kaplan, P.L. Sholnikov: Phys. Rev. Lett. **88–074801**, 1 (2002)

6.29. P.M. Paul et al.: Science **292**, 1689 (2001)

6.30. M. Hentschel et al.: Nature **414**, 509 (2001)

6.31. J. Marciak-Kozlowska, M. Kozlowski. Lasers Eng. **6**, 141 (1997)

6.32. B. Gaveau, T. Jacobson, M. Kac, L.S. Schulman: Phys. Rev. Lett. (53), 419 (1984)

6.33. S.C. Tiwari: Phys. Lett. A (133), 279 (1988)

6.34. L. de la Peña, M. Cetto: *The Quantum Dice: An Introduction to Stochastic Electrodynamics* (Kluwer 1996)

6.35. B. Haisch, A. Rueda: Phys. Lett. A **268**, (2000)

6.36. H.E. Puthoff.: Phys. Rev. **D35**, 3266 (1987)

6.37. H.E. Puthoff: Phys. Rev. **A39**, 2333 (1989)

6.38. R.P. Feynman, R.B. Leighton, M. Sands: *The Feynman Lecture on Physics* (Addison-Wesley Reading MA 1963)

6.39. T.H. Boyer: Phys. Rev. **D11**, pp 790–809 (1975)

6.40. TESLA Technical Design Report, http://tesla.desy.de/new-pages/TDR-CD/start.html

6.41. A. Zewail: J. Phys. Chem. **A104**, 5660 (2000)
6.42. M. Drescher et al.: Science **291**, 1923 (2001)
6.43. C. Wolf: Il Nuovo Cimento **102B**, 219 (1998)
6.44. H. Stumpf, Z. Natureforsch **A40**, 752 (1985)
6.45. S. Lepri, R. Livi, A. Politi: http://lanl/arXiV/cond-mat/0112193, Phys. Rep. **377**, 1 (2003)
6.46. H. Matsuda, K. Ishii: Prog. Theor. Phys. Suppl. **45**, 56 (1970)
6.47. A. Dhar: Phys. Rev. Lett. **86**, 5882 (2001)
6.48. S. Lepri, R. Livi A. Politi: Europhys. Lett. **43**, 271 (1998)
6.49. A.V. Savin et al.: Phys. Rev. Lett. **88**, 154301 (2002)
6.50. A. Dhar: Phys. Rev. Lett. **86**, 3554 (2001)
6.51. G. Casati, T. Prosen: http://lanl/arXiV/cond-mat/0203331
6.52. P. Grassberger et al.: Phys. Rev. Lett. **89**, 180601 (2002)
6.53. Chaozi Wan et al.: PNAS **97**, 14052 (2000)
6.54. J. Hone et al.: Phys. Rev. **B59**, R2514 (1999)
6.55. O. Narayan, Sz. Ramaswamy: Phys. Rev. Lett. **89**, 200601-I (2002)
6.56. H. Forsman, P. Anderson: J. Chem. Phys. **80**, 2804 (1984)
6.57. T. Fieleig et al.: Proc. Natl. Acad. Sci. USA **96**, 1187 (1999)
6.58. J.M. Lévy-Leblond, F. Balibar: *Quantons* (North Holland Physics Publishing Amsterdam 1990)

7

Faster, Brighter, Shorter

7.1 Hyperbolic Heat Transport Induced by Zeptosecond Laser Pulses

7.1.1 The LASETRON Project

In paper [7.1], was theoretically demonstrated that $10^{-21} - 10^{-22}$ s (zeptosecond and sub-zeptosecond) laser pulses can be generated using petawatt lasers, and already available terawatt lasers may generate sub-attosecond pulses of $\sim 10^{-19}$ s. The pulses will be radiated by ultrarelativistic electrons driven by circularly polarized high-intensity laser fields. They are basically reminiscent of synchrotron radiation. The major distinct feature is the forced synchronization of all radiating electrons by the driving laser field. Radiation of such a synchronized bunch would be viewed by an observer at any point in the rotation plane as huge pulses/burst of electromagnetic field as short as:

$$\Delta t \sim \frac{1}{2\omega_L \gamma^3},\qquad(7.1)$$

where γ is the electron's relativistic factor. With $\lambda_L = (2\pi c)/(\omega_L) \sim 1$ μm and $\gamma \sim 64$ (attainable with a petawatt laser), $\Delta t \sim 10^{-21}$ s. Such a system can be called *LASETRON* [7.1]. It can be achieved by placing a solid particle or a piece of wire of subwavelength cross section in the focal plane of a superenergetic laser.

The *LASETRON* as a source of the zeptosecond laser pulses opens the new possibility of the investigation of the superfast thermal processes. In this section, we discuss the structure of the Heaviside equation for the description of the interaction of the *zs* laser pulses with matter.

In addition, we investigate the generalized Schrödinger equation, which structure changes (in comparison to Schrödinger equation) for zeptosecond time scale. Both equations contain additional term proportional to the relaxation time.

The models of thermal processes with $\tau \neq 0$ we will call the causal models of the thermal phenomena in opposition to noncausal models with $\tau = 0$. The physical background for the differentiation of the causal and noncausal models stems from the observation that for $\tau = 0$, velocity of the propagation of the interaction $v \to \infty$, and $v/c \to \infty$ ($c =$ light velocity) in complete disagreement with special relativity theory.

7.1.2 Heat Transport in Zeptosecond (10^{-21} s) Time Scale

Ultrashort electromagnetic pulses have always been a great interest, largely as a means of investigating and controlling ever faster processes on different scales: molecular, atomic, and recently nuclear. Recent proposals [7.1] explored various method to attaining the shortest sub-femtosecond ($10^{-16} - 10^{-17}$) laser pulses of atomic time-scale duration. In the most recent breakthrough work [7.2], the train of 0.25 fs pulses have been observed experimentally. Time-resolved measurement with these pulses is able to trace dynamics of molecular structure but fails to capture electronic processes occurring on an attosecond time scale. Very recently [7.3], M. Hentschel et al. traced electronic dynamics with of ≤ 150 as (10^{-18} s) by using a sub-femtosecond soft X-ray pulse and a few-cycle visible light pulse. The results presented in paper [7.3] indicate an attosecond response of the atomic system, a soft X-ray pulse duration of 650 ± 150 as, and an attosecond synchronism of the soft X-ray pulse with the light field. The demonstrated experimental tools and techniques open the door to attosecond spectroscopy of bound electrons.

In the following, we propose the relativistic Heaviside equation for the study of the thermal process induced by laser pulses with time duration $\Delta t < \tau$ where τ is the characteristic relaxation time:

$$\frac{1}{v^2}\frac{\partial^2 T}{\partial t^2} + \frac{m}{\hbar}\frac{\partial T}{\partial t} + \frac{2Vm}{\hbar^2}T - \nabla^2 T = 0, \qquad (7.2)$$

$$v = \alpha c, \qquad \tau = \frac{\hbar}{m\alpha^2 c^2}.$$

In equation (7.2), m is the heat carrier mass and α is strength constant for electromagnetic ($\alpha = 1/(137)$) or strong interaction ($\alpha = 0.15$), V is the potential energy, and c is velocity of light.

In paper [7.4], it was shown that Heaviside diffusion equation (7.2) can be obtained within the frame of the correlated random walk (CRW) of the Brownian motion. As was shown in paper [7.4], the average displacement of the Brownian particle (for $V = 0$) is described by formula:

$$< x^2 >= \frac{2\hbar\tau}{m} \left[\frac{t}{\tau} - \left(1 - e^{-\frac{t}{\tau}}\right) \right] . \tag{7.3}$$

For $t > \tau$, formula (7.3) gives

$$< x^2 >= \frac{2\hbar}{m} t . \tag{7.4}$$

Formula (7.4) describes the quantum diffusion of heat carriers (*heatons*) with quantum coefficient $D = \hbar/m$.

For $t \leq \tau$, one obtains from formula (7.3)

$$< x^2 >= \alpha^2 c^2 t^2 . \tag{7.5}$$

Formula (7.5) describes the *ballistic* motion of a *heaton* with velocity $v \sim \alpha c$. For $t < \tau$ (and $V = 0$) equation (7.2), has the form of the wave equation:

$$\frac{1}{v^2} \frac{\partial^2 T}{\partial t^2} = \nabla^2 T . \tag{7.6}$$

The maximum value of the thermal wave velocity can be equal to c, the velocity of light. In that case, the relaxation time is described by formula:

$$\tau = \frac{\hbar}{mc^2} = \frac{\Lambda}{c} , \tag{7.7}$$

where Λ denotes the reduced Compton wavelength, e.g. for electrons

$$\Lambda = \frac{\hbar}{mc} . \tag{7.8}$$

It is well-known from quantum electrodynamics that quantum void fluctuations create and destroy virtual electron-positron pairs. These virtual electron-positron pairs have a characteristic lifetime of the order $\Lambda/c \sim 10^{-21}$ s $= 1$ zs. The Compton wavelength Λ can be identified as the new mean free path because it is the typical distance covered by the virtual pair before its annihilation.

It is important to take into account the fact that for $\Delta t < \tau$ zs, the *heatons* move with the velocity of light. It is worthwhile to realize that there exist models of elementary particles in which it is assumed that the electron propagates with the speed of light with certain chirality, except that at random times it flips both the direction of propagation (by 180^0) and handedness, the rate of such flips is precisely the mass m (in units $\hbar = c = 1$) [7.5], [7.6]. In the frame of the model developed in the current paper, the relaxation time for the interaction of *heatons* with voids is described by the formula $\tau = \Lambda/c = 1$ zs (for electrons).

In paper [7.7], the generalized Schrödinger equation was formulated:

$$i\hbar\frac{\partial\Psi}{\partial t} = -\frac{\hbar^2}{2m}\nabla^2\Psi + V\Psi - \frac{2\hbar^2}{m\alpha^2c^2}\frac{\partial^2\Psi}{\partial t^2} . \qquad (7.9)$$

Equation (7.9) can be written as:

$$i\hbar\frac{\partial\Psi}{\partial t} = -\frac{\hbar^2}{2m}\nabla^2\Psi + V\Psi - \frac{2\hbar}{\alpha^2}\frac{\Lambda}{c}\frac{\partial^2\Psi}{\partial t^2} . \qquad (7.10)$$

In equation (7.10), the last term proportional to Λ/c:

$$-\frac{2\hbar}{\alpha^2}\frac{\Lambda}{c}\frac{\partial^2\Psi}{\partial t^2} . \qquad (7.11)$$

describes the interaction of electrons with virtual positron-electron pairs in the void. This interaction can be investigated only with zeptosecond laser pulses emitted by *LASETRON*. In order to obtain Schrödinger equation:

$$i\hbar\frac{\partial\Psi}{\partial t} = -\frac{\hbar^2}{2m}\nabla^2\Psi + V\Psi , \qquad (7.12)$$

we are forced to assume $\Lambda/c \to 0$, i.e. $c \to \infty$ (nonrelativistic approximation) or $\Lambda \to 0$, i.e. we exclude the change of photon wavelength in Compton experiment. Both assumptions are not in agreement with experiment.

Physicists and engineers can now take snapshots of evolving atomic systems with pinpoint accuracy using femtosecond laser pulses to both trigger the dynamics and illuminate the system. This can be done by splitting each laser pulse with a partially transmitting mirror and delaying the less energetic "probe" pulse with respect to the stronger excitation or "pump" pulse. In this way, a powerful femtosecond laser pulse can initiate the same microscopic process in millions of molecules or sites in a crystal lattice. A weaker portion of the pulse (or a frequency-shifted replica) can then probe the dynamics by measuring changes in the optical properties of the system at a later instant.

It is possible to replay the atomic or molecular dynamics in slow motion by using a series of femtosecond pulses and increasing the delay between successive pump and probe pulses. This method, pump-probe spectroscopy, allows the study of microscopic dynamics. The time resolution is only limited by the duration of the pump and probe pulses.

The pulses shorter than 10 fs can be routinely generated using self-mode locking lasers. Femtosecond laser pulses allow physicists and chemists to follow these femtosecond processes by tracing the displacements of atoms, as demonstrated for the first time in the late 1980s by Ahmed Zewail [7.8]. These displacements include changes in the optical properties of the weakly bound valence electrons that can be revealed instantly by visible-light probe. In complex systems, however, the atomic motion can be more accurately determined by studying core electrons close to the atomic nucleus, but this requires X-ray wavelengths.

To the study processes inside atoms, the sub-femtosecond laser pulses are needed. In paper [7.9], the results of the production of the single soft X-ray pulses by high-order harmonic generation with 7-femtosecond (fs) laser pulses are presented. The techniques developed in paper [7.9] offer the potential for generating and measuring single attosecond X-ray pulses. Bohr's simple model of the hydrogen atom predicts that the electron takes about 140 attoseconds (as) to orbit around the proton:

$$T_{orbit} = \frac{2\pi a_{Bohr}}{v_{orbit}} = \frac{2\pi a_{Bohr}}{\alpha c} = 140 \, as \,. \tag{7.13}$$

In formula (7.13), a_{Bohr} is the Bohr radius of the atom, $a_{Bohr} = 0.5 \, 10^{-10}$ m, α is the fine structure constant for electromagnetic interaction $\alpha = 137^{-1}$, and c is the vacuum light velocity.

The thermal processes generated inside the atom with attosecond laser pulses were investigated in paper [7.4]. It was shown that for attosecond laser pulses, the master thermal equation has the form of hyperbolic quantum heat transport equation (QHT):

$$\tau \frac{\partial^2 T}{\partial t^2} + \frac{\partial T}{\partial t} = \frac{\hbar}{m_e} \nabla^2 T \,, \tag{7.14}$$

where T is the temperature, τ is the relaxation time, and m_e denotes the electron mass. The relaxation time is defined as follows

$$\tau = \frac{\hbar}{m_e \alpha^2 c^2} \tag{7.15}$$

and is of the order 10 as for the hydrogen atom. The quantum heat transport equation (QHT) describes the quantum limit of heat transport when the pulse duration Δt is of the order or smaller than the relaxation time. As was shown in paper [7.4] for $\Delta t < \tau$, the QHT has the form of the quantum wave equation

$$\frac{1}{v_h^2}\frac{\partial^2 T}{\partial t^2} = \nabla^2 T\,, \tag{7.16}$$

where $v_h = \alpha c$ is the velocity of the thermal wave. The thermal wave has the wave length λ_{th}:

$$\lambda_{th} = \frac{1}{\omega}v_{th} = \tau v_{th} = \frac{\hbar}{m v_{th}} = \lambda_B\,, \tag{7.17}$$

where $\lambda_B = \hbar/p$ denotes the de Broglie wavelength. We can conclude that for laser pulses with $\Delta t < \tau$, the thermal processes inside the atoms can be visualized as the thermal waves, in complete agreement with quantum mechanics.

Quantum mechanics based on the Schrödinger equation has been remarkably successful in all realms of atoms, molecules, and solids. In the sequel, we develop and solve the modified Schrödinger equation that describes the structure of matter as observed by attosecond laser pulses, i.e. for time periods shorter than the characteristic atomic relaxation $\tau \sim 10$ as.

The modified Schrödinger equation MSE (7.9) can be written as:

$$i\hbar\frac{\partial \Psi}{\partial t} = -\frac{\hbar^2}{2m_e}\nabla^2\Psi + V\Psi - 2\tau\hbar\frac{\partial^2\Psi}{\partial t^2}\,, \tag{7.18}$$

where the relaxation time τ is described by formula (7.15). For equation (7.18), the probability density ϱ fulfils the continuity equation [7.11]

$$\frac{\partial \varrho}{\partial t} + \text{div}\boldsymbol{S} = 0\,, \tag{7.19}$$

where ϱ equals

$$\varrho = \Psi\Psi^* - 2i\tau\left(\Psi^*\frac{\partial\Psi}{\partial t} - \Psi\frac{\partial\Psi^*}{\partial t}\right) \tag{7.20}$$

and $\boldsymbol{S}$ is the probability flow vector. By restricting the solution space we may keep $\varrho \geq 0$ [7.12].

The eigenvalue problem for one-dimensional MSE

$$i\hbar\frac{\partial\Psi(x,t)}{\partial t} = -\frac{\hbar^2}{2m_e}\frac{\partial^2\Psi}{\partial x^2} + V(x)\Psi(x,t) - 2\tau\hbar\frac{\partial^2\Psi}{\partial t^2} \tag{7.21}$$

gives for $\Psi(x,t)$ in the form:

$$\Psi(x,t) = u(x)\varphi(t), \tag{7.22}$$

$$-\frac{\hbar^2}{2m_e}\frac{\mathrm{d}^2u(x)}{\mathrm{d}x^2} + V(x)u(x) = Eu(x). \tag{7.23}$$

For time-dependent function $\varphi(t)$ we obtain the equation

$$\frac{\mathrm{d}^2\varphi(t)}{\mathrm{d}t^2} + \frac{i}{2\tau}\frac{\mathrm{d}\varphi}{\mathrm{d}t} - \frac{E}{2\tau\hbar}\varphi(t) = 0. \tag{7.24}$$

The solutions of (7.24) have the form:

$$\varphi_1(t) = \mathrm{e}^{-\frac{E_i}{\hbar}t}, \qquad \varphi_2(t) = \mathrm{e}^{i\left(\frac{Et}{\hbar} - \frac{1}{2\tau}t\right)}. \tag{7.25}$$

Considering formulae (7.22) and (7.25), one obtains the general solution of the MSE in the form:

$$\Psi(x,t) = u(x)\mathrm{e}^{-\frac{it}{2\tau}}\left[A\mathrm{e}^{-\left(\frac{Et}{\hbar} - \frac{t}{2\tau}\right)} + B\mathrm{e}^{\frac{iE}{\hbar}t}\right]. \tag{7.26}$$

The additional term (in comparison with Schrödinger equation), $\exp(-\frac{it}{2\tau})$ describes the interaction of the electron with its surrounding in atom, i.e. frictional force. It is worthwhile to recognize that this real frictional force is described by the real relaxation time (7.15). It can be imagined that in the atoms electron is moving in highly viscous medium, which is not observable for long time pulses, $\Delta t \gg \tau$.

Considering the formula (7.26), which describes the wave function for stationary state of the electron as "observed" by attosecond laser pulse, the following *scenario* of the electron motion can be formulated. In hydrogen-like atoms due to frictional force, electron moves with constant velocity $v = \alpha c$. The background viscous medium can be discovered and observed only with attosecond laser pulses.

For obtaining the Schrödinger equation, we are forced to assure $\tau \to 0$ in MSE. Considering formula (7.15), $\tau = 0$, means for $m \neq 0$, $c \to \infty$. This is in accord with the nonrelativistic nature of the Schrödinger equation. It means that Schrödinger equation allows for propagation of the information with infinite velocity. On the other hand with MSE, (7.18), the information can be transferred with finite velocity $v = \alpha c$, $v < c$.

7.1.3 Nuclear Physics with Lasers

The power in a laser pulse is defined to be energy in the pulse divided by its time duration. Clearly for the highest powers, we need high pulse energies in short pulses.

Table 7.1. The Results for the Reactions: (γ, e), (γ, n), $(\gamma, \text{Fission})$

	Duration of the Pulse fs	Power Density W/cm^2	Target Energy max	Particles	References
1	300	$\sim 10^{19}$	CH 3000Å	Electrons up to 20 MeV	G. Malka et al.: Phys. Rev. Lett. **79**, 2053 (1997)
2	300	$3.5 \cdot 10^{19}$	CD_2	Neutrons 2.5 MeV	L. Disdier et al.: Phys. Rev. Lett. **82**, 1454 (1999)
3	450	10^{20}	U^{238}	Fission fragment	T.E. Cowan: PDL **84**, 903 (2000)
4	10^3	10^{19}	U^{238}	Fission fragment	KWD Ledinghan: PDL **84**, 899 (2000)
5	35	10^{16}	D_2 Gas jet	Neutrons 2.5 MeV	T. Ditmire et al.: Nature **398**, 489 (1999)
6	500	10^{19}	CD_2	neutrons 2.5 MeV	R. Kodama et al.: Nature **412**, 798 (2001)

The invention of the technique of chirped pulse amplification (CPA) has opened new possibilities for laser science [7.13]. It has resulted in a dramatic increase in the focused intensity on the target. Intensities of $3 \times 10^{20} \mathrm{W/cm}^2$ have been demonstrated with the NOVA petawatt laser at the Lawrence Livermore National Laboratory in the USA [7.15]. Other laboratories have demonstrated intensities on the target of a few 10^{19} W/cm^2. These intensities are well into the relativistic regime. These lasers include the new laser in France [7.16], the GEKKO XII lasers in Japan [7.17], and the VULCAN laser in the United Kingdom [7.18]. In Table 7.1, the data concerning the results of the investigation of (γ, e) [7.20], (γ, n) [7.21], $(\gamma, \text{fission})$ [7.22] performed in the existing facilities are presented.

7.1.4 Energy Spectra of the Relativistic Electrons

Recently, it has become possible to produce MeV electrons with short-pulse multi-terrawatt laser system [7.16]. The fast igniter concept [7.25, 7.26] relevant to the inertial confinement fusion enhances the interest in this process. In an underdense plasma, electrons and ions tend to be expelled from the focal spot by the ponderomotive pressure of an intense laser pulse, and the formed channel [7.27, 7.28] can act as a propagation guide for the laser beam. Depending on the quality of the laser beam, the cumulative effects of ponderomotive and relativistic self-focusing [7.28] can significantly increase the laser intensity. For these laser pulses, the laser electric and magnetic fields reach few hundreds of GV/m and megagauss, respectively, and quiver velocity in the laser field is close to light speed. The component of the resulting Lorentz force $(-ev \, x \, \boldsymbol{B})$ accelerates electrons in the longitudinal direction, and energies of several tens of MeV can be achieved [7.29]. Recently the spectra of hot electrons (i.e. with energy in MeV region) were investigated. In paper [7.20], the interaction of 500 fs FWHM pulses with CH target was measured. The electrons with energy up to 20 MeV were observed. Moreover for electrons with energies higher than 5 MeV, the change of electron temperature was observed: from 1 MeV (for energy of electrons < 5 MeV) to 3 MeV (for energy of electrons > 5 MeV). In this section the interaction of femtosecond laser pulse with electron plasma will be investigated. Within the theoretical framework of Heaviside temperature wave equation, the heating process of the plasma will be described. It will be shown that in vicinity of energy of 5 MeV, the sound velocity in plasma reaches the value $\frac{c}{\sqrt{3}}$ and is independent of the electron energy.

The mathematical form of the hyperbolic quantum heat transport was proposed in [7.31] and [7.32]. Under the absence of heat or mass sources, the equations can be written as the Heaviside equations:

$$\frac{1}{v_\varrho^2}\frac{\partial^2 \varrho}{\partial t^2} + \frac{1}{D_\varrho}\frac{\partial \varrho}{\partial t} = \frac{\partial^2 \varrho}{\partial x^2} \tag{7.27}$$

and

$$\frac{1}{v_T^2}\frac{\partial^2 T}{\partial t^2} + \frac{1}{D_T}\frac{\partial T}{\partial t} = \frac{\partial^2 T}{\partial x^2} \tag{7.28}$$

for mass and thermal energy transport, respectively. The discussion of the properties of (7.27) was performed in [7.31] and (7.28) in [7.32]. In (7.27), v_ϱ is the velocity of density wave, and D_ϱ is the diffusion coefficient for mass transfer. In (7.28), v_T is the velocity for thermal energy propagation and D_T is the thermal diffusion coefficient.

In the subsequent, we will discuss the complex transport phenomena, i.e. diffusion and convection in the external field. The current density in the case when the diffusion and convection are taken into account can be written as:

$$j = -D_\varrho \frac{\partial \varrho}{\partial t} - \tau \frac{\partial j}{\partial t} + \varrho V . \tag{7.29}$$

In equation (7.29), the first term describes the Fourier diffusion, the second term is the Maxwell-Cattaneo term, and the third term describes the convection with velocity V. The continuity equation for the transport phenomena has the form:

$$\frac{\partial j}{\partial x} + \frac{\partial \varrho}{\partial t} = 0 . \tag{7.30}$$

Considering both equations (7.29) and (7.30), one obtains the transport equation:

$$\frac{\partial \varrho}{\partial t} = -\tau_\varrho \frac{\partial^2 \varrho}{\partial t^2} + D_\varrho \frac{\partial^2 \varrho}{\partial x^2} - V\frac{\partial \varrho}{\partial x} . \tag{7.31}$$

In equation (7.31), τ_ϱ denotes the relaxation time for transport phenomena. Let us perform the Smoluchowski transformation for $\varrho(x,t)$

$$\varrho = \exp\left[\frac{Vx}{2D} - \frac{V^2 t}{4D}\right] \varrho_1(x,t) . \tag{7.32}$$

After substituting $\varrho(x,t)$ formula (7.32) to equation (7.31) one obtains for $\varrho_1(x,t)$:

$$\tau_{\varrho}\frac{\partial^2 \varrho_1}{\partial t^2} + \left(1 - \tau_{\varrho}\frac{V_{\varrho}^2}{2D_{\varrho}}\right)\frac{\partial \varrho_1}{\partial t} + \tau_{\varrho}\frac{V^4}{16D_{\varrho}^2}\varrho_1 = D_{\varrho}\frac{\partial^2 \varrho_1}{\partial x^2}. \tag{7.33}$$

Considering that $D_{\varrho} = \tau_{\varrho}v_{\varrho}^2$ (7.33) can be written as

$$\tau_{\varrho}\frac{\partial^2 \varrho_1}{\partial t^2} + \left(1 - \frac{V_{\varrho}^2}{2v_{\varrho}^2}\right)\frac{\partial \varrho_1}{\partial t} + \frac{1}{16\tau_{\varrho}}\frac{V^4}{v_{\varrho}^4}\varrho_1 = D_{\varrho}\frac{\partial^2 \varrho_1}{\partial x^2}. \tag{7.34}$$

In the same manner, equation for the temperature field can be obtained:

$$\tau_T\frac{\partial^2 T_1}{\partial t^2} + \left(1 - \frac{V_T^2}{2v_T^2}\right)\frac{\partial T_1}{\partial t} + \frac{1}{16\tau_T}\frac{V_T^4}{v_T^4}T_1 = D_T\frac{\partial^2 T_1}{\partial x^2}. \tag{7.35}$$

In equation (7.35), τ_T, D_T, and V_T and v_T are relaxation time for heat transfer, diffusion coefficient, heat convection velocity, and thermal wave velocity.

Now we will investigate the structure and solution of the equation (7.35). For the hyperbolic heat transport equation (7.35), we seek a solution of the form:

$$T_1(x,t) = e^{-\frac{t}{2\tau_T}} u(x,t). \tag{7.36}$$

After substitution of (7.36) into (7.35), one obtains:

$$\tau_T\frac{\partial^2 u(x,t)}{\partial t^2} - D_T\frac{\partial^2 u(x,t)}{\partial x^2} + \left(-\frac{1}{4\tau_T} + \frac{V_T^2}{4D_T} + \tau_T\frac{V_T^4}{16D_T^2}\right)u(x,t)$$

$$- \tau_T\frac{V_T^2}{2D_T}\frac{\partial u(x,t)}{\partial t} = 0. \tag{7.37}$$

Considering that $D_T = \tau_T v_T^2$ (7.37) can be written as

$$\tau_T\frac{\partial^2 u}{\partial t^2} - \tau_T v_T^2\frac{\partial^2 u(x,t)}{\partial x^2}, \tag{7.38}$$

$$+ \left(-\frac{1}{4\tau_T} + \frac{V_T^2}{4\tau_T v_T^2} + \tau_T\frac{V_T^4}{16\tau_T^2 v_T^4}\right)u(x,t) - \frac{V_T^2}{2v_T^2}\frac{\partial u}{\partial t} = 0.$$

After omitting the term $(V_T^4)/(v_T^4)$ in comparison with the term $(V_T^2)/(v_T^2)$, (7.39) takes the form:

$$\frac{\partial^2 u}{\partial t^2} - v_T^2\frac{\partial^2 u}{\partial x^2} + \frac{1}{4\tau_T^2}\left(-1 + \frac{V_T^2}{v_T^2}\right)u(x,t) - \frac{V_T^2}{2v_T^2\tau_T}\frac{\partial u}{\partial t} = 0. \tag{7.39}$$

Considering that $\tau_T^{-2} \gg \tau_T^{-1}$ one obtains from (7.39)

$$\frac{\partial^2 u}{\partial t^2} - v_T^2\frac{\partial^2 u}{\partial x^2} + \frac{1}{4\tau_T^2}\left(-1 + \frac{V_T^2}{v_T^2}\right)u = 0. \tag{7.40}$$

Equation (7.40) is the master equation for heat transfer induced by ultra-short laser pulses, i.e. when $\Delta t \approx \tau_T$. In the following, we will consider the equation (7.40) in the form:

$$\frac{\partial^2 u}{\partial t^2} - v_T^2 \frac{\partial^2 u}{\partial x^2} - qu = 0 \tag{7.41}$$

where

$$q = \frac{1}{4\tau_T^2} \left(\frac{V_T^2}{v_T^2} - 1 \right). \tag{7.42}$$

In equation (7.42), the ratio

$$M_T = \frac{V_T}{v_T} = \frac{V_T}{v_S} \tag{7.43}$$

is the Mach number for thermal processes, for $v_T = v_S$ is the sound velocity in the gas of heat carriers [7.32].

In paper [7.32], the structure of equation (7.41) was investigated. It was shown that for $q < 0$, i.e. $V_T < v_S$, subsonic heat transfer is described by the modified telegraph equation

$$\frac{1}{v_T^2} \frac{\partial^2 u}{\partial t^2} - \frac{\partial^2 u}{\partial x^2} + \frac{1}{4\tau_T^2 v_T^2} \left(\frac{V_T^2}{v_S^2} - 1 \right) u = 0. \tag{7.44}$$

For $q > 0$, $v_S < V_T$, i.e. for supersonic case heat, transport is described by the Klein-Gordon equation:

$$\frac{1}{v_T^2} \frac{\partial^2 u}{\partial t^2} - \frac{\partial^2 u}{\partial x^2} + \frac{1}{4\tau_T^2 v_T^2} \left(\frac{V_T^2}{v_S^2} - 1 \right) u = 0. \tag{7.45}$$

The velocity of sound v_S depends on the temperature of the heat carriers. The general formula for sound velocity reads [7.30]:

$$v_S^2 = \left(zG - \frac{G}{z} \left(1 + \frac{5G}{z} - G^2 \right)^{-1} \right)^{-1}. \tag{7.46}$$

In formula (7.46), $z = (mc^2)/T$ and G is of the form [7.30]:

$$G = \frac{K_3(z)}{K_2(z)}, \tag{7.47}$$

where c is the light velocity, m is the mass of heat carrier, T is the temperature of the gas, and $K_3(z)$, $K_2(z)$ are modified Bessel functions of the second kind. In Fig. 7.1, the ratio of $\left(\frac{v_S}{c} \right)^2$ was presented as the function of $\frac{T}{mc^2}$. Figure 7.1(a) presents the $\left(\frac{u}{c} \right)^2$ for $\frac{T}{mc^2} < 1$ (nonrelativistic approximations) and Fig. 7.1(b) presents the $\left(\frac{u}{c} \right)^2$ for the very high temperature heat carriers

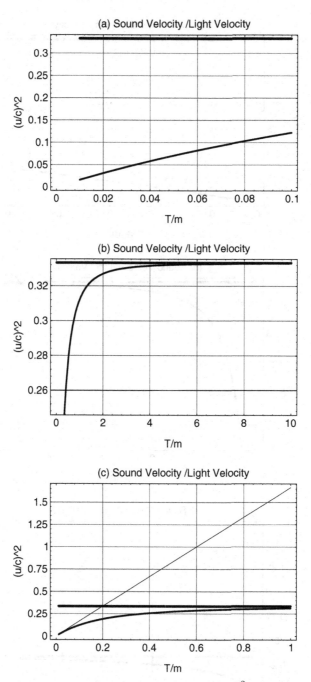

Fig. 7.1. (a) The ratio: sound velocity/light velocity $\left(\frac{u}{c}\right)^2$ as the function $\frac{T}{m}$ for cold heat carriers $\left(\frac{T}{m} \ll 1\right)$; (b) for hot heat carriers $\left(\frac{T}{m} \gg 1\right)$; (c) comparison of the ratio $\left(\frac{u}{c}\right)^2$ for hot carriers (—), ultrarelativistic carriers (—), and Maxwellian approximation (—).

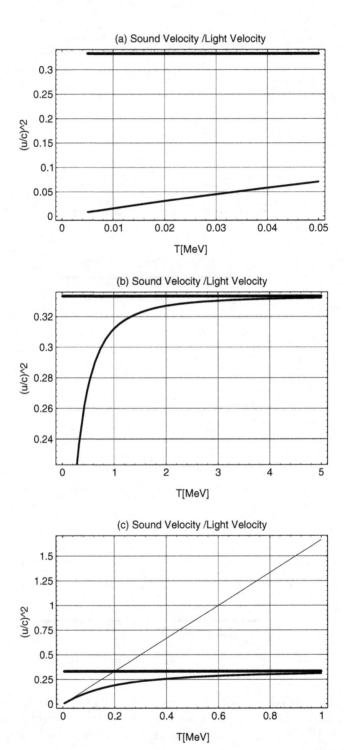

Fig. 7.2. The same as in Fig. 7.1 but for electrons with mass $m = 0.51$ MeV/c^2.

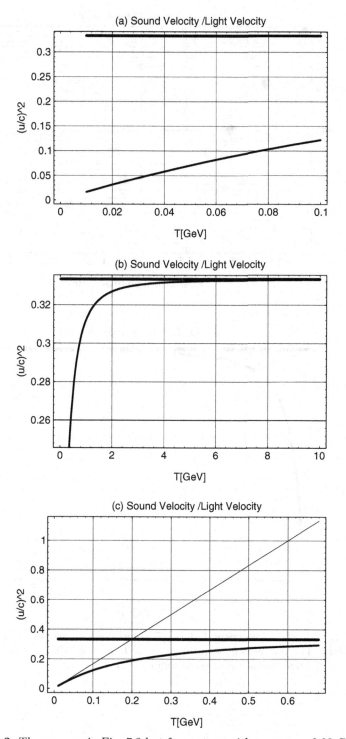

Fig. 7.3. The same as in Fig. 7.2 but for protons with mass $m = 0.98$ GeV.

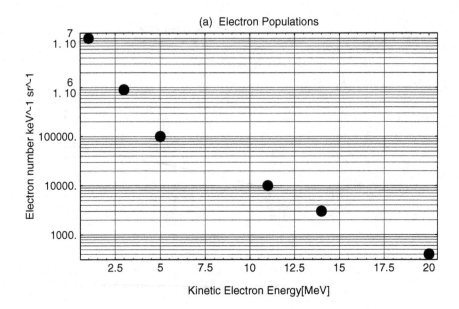

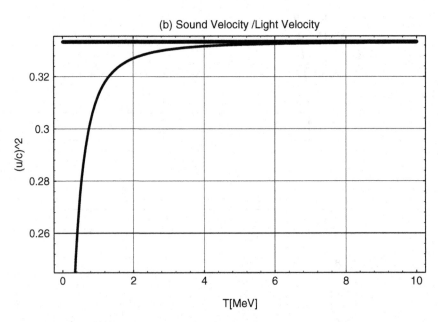

Fig. 7.4. (**a**) The experimental data [7.7] for the electron populations. Circles = data. The interaction beam intensity is $I_{\max} \approx 6 \cdot 10^{18}$ W/cm^2. (**b**) The ratio $\frac{u}{c}$ as the function of the electron temperature T [MeV].

gas, i.e. $T > mc^2$ (relativistic gas). It is interesting to observe that for non-relativistic gas, v_S^2 is a linear function of temperature. From formula (7.47), it can be concluded [7.30] that for $T < mc^2$ one obtains

$$\left(\frac{v_S}{c}\right)^2 = \left(\frac{5T}{3mc^2}\right) \tag{7.48}$$

i.e as for Maxwellian nonrelativistic gas. On the other hand for $T \gg mc^2$, $\left(\frac{v_S}{c}\right)^2 = \frac{1}{3}$ and is independent of T where $v_S^2 = \frac{c^2}{3}$ denotes the sound velocity in the photon gas. In the sequel, we will study the heat transfer in the relativistic gas, i.e. when $\left(\frac{v_S}{c}\right)^2$ is constant. In that case, (7.44) and (7.45) are the linear hyperbolic equations. In Fig. 7.1(c), it is shown that Maxwellian approximation is not valid for $\frac{T}{mc^2} > 0.05$ and moreover gives a wrong description of v_S for $\frac{T}{mc^2} = 0.6$, for $v_S > c$ in complete disagreement with special relativity theory. In Fig. 7.2(a), (b), and (c) the results of calculations of the sound velocity of electron gas and in Fig. 7.3(a), (b), and (c) for proton gas are presented, respectively. For the initial conditions

$$u|_{t=0} = f(x), \qquad \frac{\partial u}{\partial t}\bigg|_{t=0} = F(x).$$

Solutions of the equation can be found in [7.33]:

$$u(x,t) = \frac{f(x - v_T t) + f(x + v_T t)}{2} + \frac{1}{2}\int_{x - v_T t}^{x + v_T t} \Phi(x, t, z)dz, \tag{7.49}$$

where

$$\Phi(x, t, z) = \frac{1}{v_T}F(z)J_0\left(\frac{\sqrt{q}}{v_T}\sqrt{(z - x)^2 - v_T^2 t_2}\right)$$

$$+ \sqrt{q}tf(z)\frac{J_0'\left(\frac{\sqrt{q}}{v_T}\sqrt{(z - x)^2 - v_T^2 t^2}\right)}{\sqrt{(z - x)^2 - v_T^2 t^2}}$$

and

$$q = \frac{1}{4\tau_T^2}\left(\frac{V_T^2}{vT2} - 1\right).$$

The general equation for complex heat transfer: diffusion plus convection can be written as:

$$\frac{\partial T}{\partial t} = -\tau_T\frac{\partial^2 T}{\partial t^2} + D_T\frac{\partial^2 T}{\partial x^2} - V_T\frac{\partial T}{\partial x}. \tag{7.50}$$

Considering (7.32), (7.36), and (7.49), the solution of equation (7.50) is

$$T(x,t) = \exp\left[\frac{V_T x}{2D} - \frac{V_T^2 t}{4D}\right] e^{-\frac{2}{2\tau_T}} \times \left(\frac{f(x - v_T t) + f(x + v_T t)}{2}\right.$$

$$\left. + \frac{1}{2} \int\limits_{x - v_T t}^{x + v_T t} \Phi(x,t,z)\mathrm{d}z\right).$$

In Fig. 7.4, the comparison of the calculation of sound velocity for electron gas and the electron spectra [7.20] is presented. The change of the shape of the electron spectra in vicinity of 5 MeV can be easily seen. At the total energy of 5 MeV, the sound velocity in electron plasma is constant and independent of electron energy. Electrons with velocities greater than $\frac{1}{\sqrt{3}}c$ can generate the shock thermal waves that heat the plasma to higher temperature.

7.2 Possible Thermal Waves Generation by Femtosecond TESLA Free Electron Laser (FEL)

7.2.1 The FEL TESLA Project

Recently (March 2001) [7.24], the international TESLA collaboration, together with members of various study groups released the TESLA Technical Design Report. In the report, the plans for a superconducting linear electron-positron collider, TESLA, were presented. The TESLA multidisciplinear accelerator offers the possibility for the study of particle physics as well of the materials physics (solid state and nanotechnology). The X-ray free electron laser laboratory proposed as part of the TESLA project is conceived as a multi-user facility following the experience of existing large synchrotron radiation facilities like HASYLAB at DESY and ESRF in Grenoble.

X-ray plays a crucial role when the structural and electronic properties of matter are to be studied on an atomic scale - particulary when looking at atoms in molecules, in large biomolecular complexes, and in condensed matter. X-rays are one of the most important tools in basic science and medical diagnostics as well as in industrial R&D.

The TESLA free electron laser (FEL) will open up a whole range of new possibilities for X-ray research: it will provide lateral fully coherent polarized X-ray wavelengths between 1 and 0.1 nm and with peak brilliance more than a 100 million times as high as are available today from the best synchrotron radiation sources. In addition, the X-rays will be delivered in flashes with a duration of 100 fs or less allowing the observation of fast chemical processes with atomic-scale spatial resolution. The X-ray laser is expected to play an important role in the structural and functional analysis of large molecular complexes that are crucial to the functioning of cells but extremely difficult to crystalize and study using present-day techniques.

The X-ray from TESLA FEL will probe the dynamic state of matter and can thus be used to study nonequilibrium states and very fast transitions between the different states of matter.

In the paper [7.19], we build up the theoretical framework for experiments planned or that can be planned for TESLA. In the book, we consider nonequilibrium states of the matter (on nuclear, atomic, or bulk) produced by the ultrafast thermal process, i.e. when the involved time periods are of the order the relaxation time. In these cases the master equation, Heaviside equation, has the form (in one-dimensional case):

$$\frac{\partial^2 T}{\partial x^2} - \frac{1}{v^2}\frac{\partial^2 T}{\partial t^2} - \frac{m_0 \gamma}{\hbar}\frac{\partial T}{\partial t} - \frac{2V m_0 \gamma}{\hbar^2}T = 0. \qquad (7.51)$$

In equation (7.51), T is the temperature, m_0 is mass of the heat carriers, V is potential barrier, v is velocity of the interaction (electromagnetic, strong) propagation, and $\gamma = (1 - v^2)^{-1/2}$ is Lorentz factor.

In the subsequence, we present the solution of the equation (7.51) for electromagnetic and strong interactions that mediate the thermal processes for energies available at the TESLA project.

7.2.2 Solution of the Heaviside Equation

With the TESLA collider (with FEL), the energies up to 500 GeV (in CMS) and ~ 10 keV (for photons) will be available. For these energies, both electromagnetic and strong interactions mediate the reactions between building blocks of matter. For electromagnetic interactions, the Heaviside equation has the form:

$$\frac{\partial^2 T}{\partial x^2} - \frac{1}{v_{\text{em}}^2}\frac{\partial^2 T}{\partial t^2} - \frac{m_0 \gamma}{\hbar}\frac{\partial T}{\partial t} - \frac{2V m_0 \gamma_{\text{em}}}{\hbar^2}T = 0, \qquad (7.52)$$

where

$$v_{\text{em}} = \alpha_{\text{em}}c; \qquad \gamma_{\text{em}} = \frac{1}{\sqrt{1 - \alpha_{\text{em}}^2}} \qquad (7.53)$$

and

$$\alpha_{\text{em}} = \frac{1}{137}.$$

Analogously for strong interactions, we obtain:

$$\frac{\partial^2 T}{\partial x^2} - \frac{1}{v_s^2}\frac{\partial^2 T}{\partial t^2} - \frac{m_0\gamma}{\hbar}\frac{\partial T}{\partial t} - \frac{2V m_0\gamma_s}{\hbar^2}T = 0 \tag{7.54}$$

and

$$\alpha_s = 0.12, \qquad \gamma_s = \frac{1}{\sqrt{1-\alpha_s^2}}. \tag{7.55}$$

Setting in equations (7.52), (7.54):

$$T^{\mathrm{em,s}} = f^{\mathrm{em,s}}(t)v^{\mathrm{em,s}}(x,t) \tag{7.56}$$

we shall show that this change of variables yields a new equation with no first derivative terms. Performing the appropriate differentiations on (7.56) and using equations (7.52), (7.54) yields

$$f\frac{\partial^2 v}{\partial x^2} - v^{-2}f\frac{\partial^2 v}{\partial t^2} - \left(2v^{-2}\frac{\mathrm{d}f}{\mathrm{d}t} + \frac{m_0\gamma}{\hbar}f\right)\frac{\partial v}{\partial t}$$

$$- \left(v^{-2}\frac{\mathrm{d}^2 f}{\mathrm{d}t^2} + \frac{m_0\gamma}{\hbar}\frac{\mathrm{d}f}{\mathrm{d}t} + \frac{2V_0 m_0\gamma}{\hbar^2}f\right)v = 0, \tag{7.57}$$

where $v = v^{\mathrm{em,s}}$, $\gamma = \gamma^{\mathrm{em,s}}$.

Thus $f(t)$ is arbitrary, so that we may choose it such that the coefficient of the $\mathrm{d}f/\mathrm{d}t$ term vanishes. This gives $2v^{-2}\frac{\mathrm{d}f}{\mathrm{d}t} + \frac{m_0\gamma}{\hbar}f = 0$, which yields

$$f(t) = e^{-\frac{m_0\gamma v^2}{2\hbar}t}, \tag{7.58}$$

where we can define the characteristic relaxation time

$$\tau_{\mathrm{em,s}} = \frac{\hbar}{m_0\gamma_{\mathrm{em,s}}v_{\mathrm{em,s}}}. \tag{7.59}$$

Then the PDE (7.57) for $v(x,t)$ has no first derivative term and reduces to

$$\frac{\partial^2 v}{\partial x^2} - \frac{1}{v^2}\frac{\partial^2 v}{\partial t^2} + qv = 0, \tag{7.60}$$

$$q = \frac{2V m_0\gamma}{\hbar^2} - \frac{m_0^2\gamma^2 v^2}{4\hbar^2},$$

Depending on the sign of q, Heaviside equation is: for $q > 0$ Klein-Gordon equation, and modified telegraph equation for $q < 0$. For $q = 0$, equation (7.60) is the wave equation with the solution:

$$u(x,t) = e^{-\frac{t}{2\tau}}\left[F(x-vt) + G(x+vt)\right], \tag{7.61}$$

where $F(x - vt)$ represents a progressing wave while $G(x + vt)$ represents the regressing wave. The waves travel undistorted but damped. These waves are therefore called *relatively undistorted*.

In the case when $q \neq 0$, we set

$$v(x, t) = F(x - \eta t) \tag{7.62}$$

$$\eta \neq v$$

and one obtains from (7.60):

$$\frac{\partial^2 F}{\partial t^2} + \frac{qv^2}{v^2 - \eta^2} F = 0. \tag{7.63}$$

Equation (7.63) tells us that we have as the solutions only sines, cosines, or hyperbolic functions as long as $v \neq \eta$. Waves of this nature (for $q \neq 0$) are *dispersive waves* because they travel with variable phase velocity. The term qv in (7.60) produces dispersion.

7.2.3 The Quantal Temperature Field $T(x, t)$

As was shown in [7.19], we can define the quantum of the temperature field the *heaton*, T_h, viz.:

$$T_h = \frac{m_0 v^2}{\sqrt{1 - \frac{v^2}{c^2}}}; \qquad v = \alpha_{\text{em,s}} c. \tag{7.64}$$

In Fig. 7.5(a), we present the *heaton* temperatures (formula (7.64)) for strong (thick line) and electromagnetic (thin line) interactions. The mass spectrum for both interactions is discrete. In the case of strong interacting particle, the threshold for thermal spectrum is equal to the "freeze-out" temperature $T_{\text{fo}} \sim 140$ MeV $= \pi$- meson mass and the particle spectrum temperature is equal:

$$T = T_{\text{fo}} + T_h. \tag{7.65}$$

For electromagnetic interactions $T_{\text{fo}} = 0$, as photon is the massless particle. In Fig. 7.5(b), we present the particle spectrum temperatures (formula (7.65)). In Fig. 7.5(c), the comparison of the theoretical calculation and measured temperature [7.10] for pions (mass $= 140$ MeV/c^2), k-meson (mass $= 490$ MeV/c^2), and nucleons ($m = 980$ MeV/c^2) is shown.

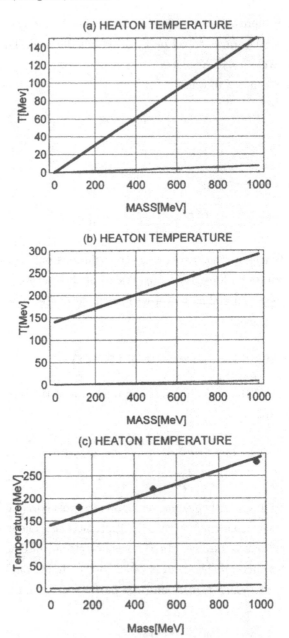

Fig. 7.5. (a) Theoretical *heaton* temperatures as the function of mass. Thick line (—) for strong interaction $\alpha_s = 0.15$, thin line (—) for electromagnetic interaction $\alpha_{em} = 1/137$. (b) *Heaton* temperature with freeze-out temperature T_{fo} = pion mass added. (c) Comparison of the theoretical and measured [7.19] *heaton* temperatures for pion, kaon, and nucleon.

For both type of interactions: electromagnetic and strong, with TESLA available energies, the damped thermal waves (equation (7.52), (7.54)) can be generated. The thermal waves represent fast, much faster than diffusion thermal processes. In the paper, we presented the first approximation of the description for the involved thermal processes – we assume the linear Heaviside equation in which the relaxation times and velocities are constant. It must be recognized, however, that in reality the coupling constant for high energy can be energy dependent (running constant).

7.3 New Laser Facility LUX (Liniac-Based Ultrafast X-ray Laser Facility)

7.3.1 The LUX Project

Ultrafast X-rays have been identified in numerous workshop and reports around the world as a key area that is ripe for new scientific investigations – with femtosecond pulses allowing the detailed study of atomic motion during physical, chemical, and biological reactions. Ultrafast lasers covering most of the visible, infrared, and ultraviolet regions of the spectrum already provide the capability to measure bond breaking in chemical reactions with both excellent timing resolution and very short pulses. Thus, experimenters have used lasers to tremendous advantage in thousands of investigations of time dynamics, many of which are absolutely critical to research in solid-state physics, semiconductors, photochemistry, and photobiology. However, ultrafast time domain studies in the X-ray region have been almost completely lacking, even though they are needed to refine the picture of dynamics at the time scales of atomic vibration periods – about 100 fs or less, and even the possibility of resolving electron dynamics with sub-femtosecond resolution.

LUX – a Liniac based ultrafast X-ray/laser facility – is a concept that is designated to produce ultrashort X-ray pulses in a highly refined manner for experiments across all areas of the physical, chemical, and biological sciences [7.23]. The facility will provide an increase of X-ray flux by several orders of magnitude and would be accessible to a large number of users. Ultrafast lasers would be available for "pump-probe" experiments at femtosecond resolution, where a pulse from a laser excites or "pumps" the system under study, while the X-ray pulse is used to probe the system configuration as a snapshot in time after the pump pulse. Figure 7.6 shows a schematic of this concept.

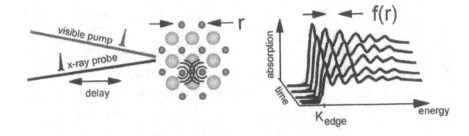

Fig. 7.6. Schematic of the LUX concept.

The LUX proposal is based on a recirculating electron liniac, which provides a compact and cost-effective configuration for the production of intense ultrafast extreme ultraviolet (EUV) and X-ray pulses, with high synchronization to sample excitation lasers. The facility is designated to produce ultrafast EUV and soft X-rays by a harmonic-cascade free-electron laser (FEL) technique while hard X-rays are produced by a novel manipulation of the electron bunches followed by compression of the photon beam.

The FEL process is initiated by lasers, which allows tunability of both wavelength and pulse duration from hundreds to tens of femtoseconds. Hard X-ray pulses are produced in superconding insertion devices – undulators produce narrow band peaks with harmonics out to 10 keV and higher, and wigglers produce broadband pulses extending to even shorter wavelengths.

Sophisticated laser systems will be an integral part of the LUX facility, providing experimental excitation pulses and stable timing signals, as well as the electron source through the photocathode laser.

7.3.2 Study of the Ultrafast Dynamics Across a Wide Range of Science

Ultrafast X-rays have been identified worldwide in numerous workshops and reports as a key area ripe for new scientific investigations. Lasers successfully cover most of the visible, infrared, and ultraviolet regions of the spectrum with both high resolution and very short pulses. Thus, experimentalists have used lasers to tremendous advantage for thousands of time dynamics investigations, many absolutely critical to scientific fields of solid-state physics, semiconductors, photochemistry, and photobiology. Existing laser-based X-ray fluxes are low, the signal levels weak, and experiments are challenging to accomplish by individual scientists. The LUX recirculating liniac-based

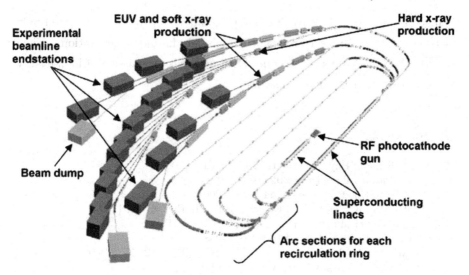

Fig. 7.7. The machine layout.

facility provides an increase of X-ray flux by several orders of magnitude, is accessible to a large number of users, with resources available for set-up of pump-probe femtosecond-scale time-resolved experiments using ultrafast lasers. The science to be carried out with LUX facility cuts across all scientific disciplines. By combining both diffraction to explore nuclear positions in real time and spectroscopy to interrogate electronic and atomic states and their structural parameters and chemical environments, the facility represents a powerful combination to address scientific problems.

Broad categories of possible experiments include:

Photoinduced phase transitions.

Metal-insulator photo-induced transitions.

Time domain structured biology.

Laser-induced continuum dressing of atoms and molecules.

Nanoparticle physics, etc.

The basic techniques of interrogation involve the pump-probe with visible laser pump light. The LUX will be designated to accommodate rapidly emerging multidimensional coherent laser spectroscopies, e.g. three laser pump beams and X-ray probe. Most of these novel forms of spectroscopies with X-rays have not even been delineated yet.

The machine layout is shown in Fig. 7.7. After acceleration to 2.3–3 GeV, the electron bunches pass through undulators to produce radiation over a range of a few tens of eV to 12 keV EUV, and soft X-rays are generated by a seeded free-electron laser process in a high-gain harmonic generation scheme (HGHG). Hard X-rays are produced by spontaneous emission of high-energy electrons in short period undulators.

References

7.1. A.E. Kaplan: Phys. Rev. Lett. **88**, 074801-1 (2002)
7.2. P.M. Paul et al.: Science **292**, 1689 (2001)
7.3. M. Heatschel et al.: Nature **414**, 509 (2001)
7.4. J. Marciak-Kozlowska, M. Kozlowski: Lasers Eng. **6**, 141 (1997)
7.5. B. Gaveau et al.: Phys. Rev. Lett. **53**, 419 (1994)
7.6. S.C. Tiwari: Phys. Lett. **A133**, 279 (1988)
7.7. J. Marciak-Kozlowska, M. Kozlowski: Lasers Eng. **12**, 53 (2002)
7.8. A. Zewail: J. Phys. Chem. **A104**, 5660 (2000)
7.9. M. Drescher et al.: Science **291**, 1923 (2001)
7.10. I.G. Bearden et al.: Phys. Rev. Lett. **78**, 2080 (1997)
7.11. C. Wolf: Il Nuovo Cimento **102B**, 219 (1998)
7.12. H. Stumpf: Z Natureforsh **A40**, 752 (1985)
7.13. K.W.D. Ledingham, P.A. Norreys: Contemporary Physics **40**, 367 (1999)
7.14. D. Stricland and G. Mouron: Optics Commun. **56**, 219 (1985)
7.15. M.D. Perry et al.: Optics Lett. **24**, 160 (1999)
7.16. N. Blanchot et al.: Optics Lett. **20**, 395 (1995)
7.17. M. Mori et al.: Nucl. Instrum. Meth. Phys. Res. **A410**, 367 (1998)
7.18. C.N. Danson et al.: J. Mod. Optics **45**, 1653 (1998)
7.19. M. Kozlowski, J. Marciak-Kozlowska: Lasers Eng. **12**, 95 (2002)
7.20. G. Malka et al.: Phys. Rev. Lett. **79**, 2053 (1997)
7.21. L. Disdier et al.: Phys. Rev. Lett. **82**, 1454 (1999) T. Ditmire et al.: Nature **398**, 489 (1999) R. Kodama et al.: Nature **412**, (2001)
7.22. T.E. Cowan: Phys. Rev. Lett. **84**, 899 (2000)
7.23. J.N. Corlet et al.: http://lux.lbl.gov
7.24. TESLA Technical Design Report, http://tesla.desy.de/new-pages/TDR-CD/start.html
7.25. M. Tabak et al.: Phys. Plasmas **1**, 1626 (1994)
7.26. P. Monot et al.: Phys. Rev. Lett. **74**, 2953 (1995)
7.27. M. Borghesi et al.: Phys. Rev. Lett. **78**, 879 (1997)
7.28. A. Borisov et al.: Plasma Phys. Controlled Fusion **37**, 569 (1995)
7.29. G. Malka et al.: Phys. Rev. Lett. **78**, 3314 (1997)
7.30. J.L. Synge: *The Relativistic Gas* (North-Holland Publisher Company 1957)
7.31. J. Marciak-Kozlowska, M. Kozlowski: Lasers Eng. **12**, 17 (2002)
7.32. J. Marciak-Kozlowska, M. Kozlowski: Lasers Eng. **14**, 125 (2004)
7.33. M. Kozlowski, J. Marciak-Kozlowska: *From Quarks to Bulk Matter*, (Hadronic Press USA 2001)

8

Fundamental Physics with Attosecond Laser Pulses

8.1 Precision Measurements of the Fundamental Constants of Nature

8.1.1 Rydberg Constant and Lamb Shift in the Hydrogen Atom

The electron and proton are involved in various phenomena in different fields of physics. As a results fundamental physical constants related to their properties (such as the electron/proton charge e, the electron and proton masses, m_e and m_p) the Rydberg constant (Ry), and the fine structure constant α, can be traced in various basic equations of the physics of atoms, molecules, solid state, nuclei, and elementary particles [8.1]. Until recently, most accurate measurements came from radio-frequency experiments only, which supplied us with precise values of most of the fundamental constants and accurate tests of the quantum theory of simple atoms.

The discovery of the Lamb shift in the hydrogen atom was a starting point of quantum electrodynamics (QED). It turned out that the two-photon Doppler-free spectroscopy of gross structure transitions (such as 1 s–2 s, 1 s–3 s, etc.) allowed access to narrower levels and could deliver very accurate values sensitive to QED effects. But to interpret those values in terms of the Lamb shift, two problems had to be solved:

- the Rydberg constant determines a dominant part of any optical transition and has to be known itself,
- a number of levels are involved and it is necessary to be able to find relationships between the Lamb shifts (E_L) of different levels.

The former problem has been solved by comparison of two transitions determining both: the Rydberg and the QED contributions. Currently, the two best results to combine are the 1 s–2 s frequency in hydrogen and deuterium [8.2, 8.3]. The latter problem has been solved with a specific difference [8.4]

$$\Delta(n) = E_{\mathrm{L}}(1\text{ s}) - n^3 E_{\mathrm{L}}(n\text{ s}) \tag{8.1}$$

which can be calculated more accurately than the Lamb shift of the individual levels.

A successful deduction of the Lamb shift, Fig. 8.1, in the hydrogen atom provides us with a precision test of bound-state QED and offers an opportunity to learn more about the proton size.

The second quantity deduced from the optical measurements on hydrogen and deuterium is the Rydberg constant. The recent progress is clearly seen from the recommended CODATA values of 1986 and 1998 [8.5].

$$Ry_{86} = 10973731.534(13)\,\mathrm{m}^{-1}, \tag{8.2}$$

$$Ry_{98} = 10973731.568549(83)\,\mathrm{m}^{-1}.$$

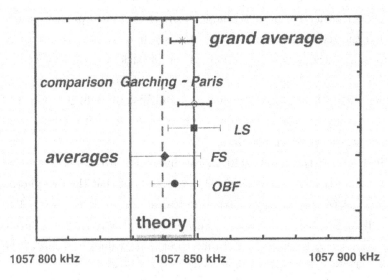

Fig. 8.1. Present status of the Lamb splitting in hydrogen ($2s_{1/2} - 2p_{1/2}$). The figure contains values derived from various experiments. *LS* stands for the Lamb splitting measurements, *FS* is for fine structure, and *OBF* stands for optical beat frequency (simultaneous measurement of two optical transitions). The theoretical estimation is taken from [8.26].

8.1.2 Zero-Point Field (ZPF) Effects on the Stability of Matter

During the 20th century, our knowledge regarding space and the properties of the vacuum has progressed considerably. In the popular meaning the vacuum is considered to be a void or "nothingness." This is the definition of a *bare vacuum*. However, with the progress of science, a new and contrasting description has arisen, which physicists call the *physical vacuum*. The *physical vacuum* contains measurable energy. This energy is called the *zero-point energy* (ZPE) because it exists even at absolute zero. The very fruitful theoretical framework in which we can describe the *zero-point energy* is the stochastic electrodynamics (SED) [8.7]–[8.10]. In the SED approach, the *physical vacuum* at the atomic or subatomic level may be considered to be inherently composed of a turbulent sea of randomly fluctuating electromagnetic field.

These fields exist at all wavelengths longer than the Planck length. At the macroscopic level, these *zero-point fields* (ZPF) are homogenous and isotropic.

The atomic building blocks of matter are dependent upon the ZPF for their very existence. This was demonstrated by H. Puthoff [8.9, 8.10]. Puthoff started by pointing out the anomaly. According to classical concepts, an electron in orbit around the proton should be radiating energy. As a consequence, as it loses energy, it should spiral into the atomic nucleus. But that does not happen. In quantum mechanics it is explained by the *Bohr's quantum conditions*. Instead of the Bohr model of the atom, Puthoff approached this problem with the assumption that the classical laws of electrodynamics were valid and that the electron is therefore losing energy and the loss was exactly balanced by energy gain from the ZPF.

In this section, we adapted the Puthoff's results to study the heat transport on the atomic level. To that aim we consider the quantum heat transport (QHT) equation [8.11]. It will be shown that at the atomic level, the structure of the QHT is dependent upon the ZPF. The condition for the quantum heat transport limit [8.11] guarantees the balance of the loss-gain energy on the atomic level. This opens a new field of investigation for laser scientists and engineers. The interaction of the ultrashort laser pulses ($\Delta t \sim$ attosecond) with matter can be used as the source of the information on the ZPF. Maybe future engineers will be specialized in "vacuum engineering."

In the stochastic electrodynamics (SED) [8.7]–[8.10], the physical vacuum is assumed to be filled with random classical zero-point electrodynamic radiation that is homogenous, isotropic, and Lorentz invariant. Writing as a sum over plane waves, the random radiation can be expressed as [8.9]

$$E^{\text{zp}}(\boldsymbol{r},t) = \text{Re} \sum_{\delta=1}^{2} \int \mathrm{d}^3 k \hat{\varepsilon} \left(\frac{\hbar\omega}{8\pi^3\varepsilon_0}\right)^{1/2} \cdot \mathrm{e}^{(\mathrm{i}\boldsymbol{kr}-\mathrm{i}\omega t+\mathrm{i}\Theta(k,\delta))}, \qquad (8.3)$$

$$H^{\text{zp}}(\boldsymbol{r},t) = \text{Re} \sum_{\delta=1}^{2} \int \mathrm{d}^3 k (\hat{k} \cdot \hat{\varepsilon}) \left(\frac{\hbar\omega}{8\pi^3\mu_0}\right)^{1/2} \cdot \mathrm{e}^{(\mathrm{i}\boldsymbol{kr}-\mathrm{i}\omega t+\mathrm{i}\Theta(k,\delta))},$$

where $\delta = 1, 2$ denote orthogonal polarizations, $\hat{\varepsilon}$ and $\hat{k}$ are orthogonal unit vectors in the direction of the electric field polarization and wave propagation, vectors, respectively, $\Theta(\boldsymbol{k},\delta)$ are random phases distributed uniformly on the interval 0 to 2π (independently distributed for each $\boldsymbol{k}, \delta$), and $\omega = kc$. It must be stressed that in the SED the *zero-point field* is treated in every way as a real, physical field.

In the subsequent, we will approximate the matter as the ensemble of the one-dimensional charged harmonic oscillators of natural frequency ω_0 immersed in the *zero-point field*. For orientation along the x axis, the (nonrelativistic) equation of motion for a particle of mass m and charge e, including damping is given by [8.9]

$$m\frac{\mathrm{d}^2 x}{\mathrm{d}t^2} + m\omega_0^2 x = \left(\frac{e^2}{6\pi\varepsilon_0 c^3}\right)\frac{\mathrm{d}^3 x}{\mathrm{d}t^3} + e e_x^{\text{zt}}(0,t), \qquad (8.4)$$

where e is the charge on electron, c is the light velocity, and ε_0 is the electrical permittivity of the vacuum.

Substitution of formula (8.3) into formula (8.4) gives the following expression for displacement and velocity:

$$x = \frac{e}{m}\text{Re} \sum_{\delta=1}^{2} \int \mathrm{d}^3 k (\hat{\varepsilon} \cdot \hat{x}) \left(\frac{\hbar\omega}{8\pi^3\varepsilon_0}\right)^{1/2} \frac{1}{D} \cdot \mathrm{e}^{(\mathrm{i}\boldsymbol{kr}-\mathrm{i}\omega t+\mathrm{i}\Theta(k,\delta))}, \qquad (8.5)$$

$$v = \frac{\mathrm{d}x}{\mathrm{d}t} = \frac{e}{m}\text{Re} \sum_{\delta=1}^{2} \int \mathrm{d}^3 k (\hat{\varepsilon} \cdot \hat{x}) \left(\frac{\hbar\omega}{8\pi^3\varepsilon_0}\right)^{1/2} \cdot \left(-\frac{\mathrm{i}\omega}{D}\right) \mathrm{e}^{(\mathrm{i}\boldsymbol{kr}-\mathrm{i}\omega t+\mathrm{i}\Theta(k,\delta))},$$

where

$$D = -\omega^2 + \omega_0^2 - \mathrm{i}\Gamma\omega^3, \qquad (8.6)$$

$$\Gamma = \frac{e^2}{6\pi\varepsilon_0 mc^3}.$$

From (8.3) and (8.5), the average power absorbed by oscillator from ZPF can be calculated [8.9], viz.:

$$\langle P^{\text{abs}} \rangle = \langle e E^{\text{zp}} \cdot \boldsymbol{v} \rangle = \frac{e^2 \hbar \omega_0^3}{12 \pi \varepsilon_0 m c^3} \, . \tag{8.7}$$

We now recognize that for "planetary" motion of electrons in the atom, the ground state circular orbit of radius r_0 constitutes a pair of one-dimensional harmonic oscillators in a plane

$$x = r_0 \cos \omega_0 t \, , \tag{8.8}$$

$$y = r_0 \sin \omega_0 t \, .$$

Therefore the power absorbed from the background by the electron in circular orbit is double of (8.7) or

$$\langle P^{\text{abs}} \rangle_{\text{circ}} = \frac{e^2 \hbar \omega_0^3}{6 \pi \varepsilon_0 m c^3} \, . \tag{8.9}$$

The power radiated by charged particle in circular orbit with acceleration A is given by the expression [8.12]

$$\langle P^{\text{rad}} \rangle_{\text{circ}} = \frac{e^2 A^2}{6 \pi \varepsilon_0 c^3} = \frac{e^2 r_0^2 \omega_0^4}{6 \pi \varepsilon_0 c^3} \, . \tag{8.10}$$

In monograph [8.11], the quantum heat transport equation for electrons in matter was formulated:

$$\frac{\lambda_B}{v_h} \frac{\partial^2 T^e}{\partial t^2} + \frac{\lambda_B}{\lambda_m} \frac{\partial T}{\partial t} = \frac{\hbar}{m_e} \nabla^2 T \, . \tag{8.11}$$

In Eq. (8.11), T is the temperature, and λ_B and λ_m are the reduced de Broglie wavelength and mean free path (for electron), respectively

$$\lambda_B = \frac{\hbar}{p}, \qquad \lambda_m = v \tau \, , \tag{8.12}$$

where v is the electron velocity and τ is the relaxation time for electrons.

In the following, we will study the quantum limit of the heat transport in the fermionic system [8.11]. We define the quantum heat transport limit as follows

$$\lambda_B = \lambda_m \, . \tag{8.13}$$

In that case, (8.11) has the form

$$\tau \frac{\partial^2 T}{\partial t^2} + \frac{\partial T}{\partial t} = \frac{\hbar}{m} \nabla^2 T \, , \tag{8.14}$$

where

$$\tau = \frac{\hbar}{m_e v^2} .$$ (8.15)

Having the relaxation time τ, one can define the pulsation ω [8.11]

$$\omega = \tau^{-1} = \frac{mv^2}{\hbar} .$$ (8.16)

For an electron in atom, $\omega = \omega_0$ (formula (8.8)), i.e.

$$\omega_0 = \frac{mv^2}{\hbar} .$$ (8.17)

Considering that for circular orbit $v = \omega_0 r_0$, formula (8.17) gives

$$r_0^2 = \frac{\hbar}{m\omega_0} .$$ (8.18)

Substituting formula (8.18) into formula (8.10), one obtains

$$\langle P^{\mathrm{rad}} \rangle_{\mathrm{circ}} = \frac{e^2 \hbar \omega_0^3}{6\pi\varepsilon_0 mc^3} = \langle P^{\mathrm{abs}} \rangle_{\mathrm{circ}} .$$ (8.19)

We conclude that in the SED framework, the QHT equation (8.14) describes the heat transport on the atomic level where the τ is the relaxation time for the electron–*zero-point field* interaction. It is quite interesting to observe that formula (8.17) is the Bohr formula for the ground state of hydrogen atom. It means that the ground state of the hydrogen atom is the result of the balance between radiation emitted due to acceleration of the electron and radiation absorbed from the zero-point background. For the first time, the balance between two forms of radiation in hydrogen atom was hypothesized by Boyer [8.13].

8.2 The Life of the Universe

8.2.1 Aging of the Universe and the Fine Structure Constant

The distant future of the Universe is dramatically different depending on whether it expands forever or stops expanding at some future time and recollapses. The long-term future of life and civilization has been discussed by J.N. Islam [8.14] and F.J. Dyson [8.15]. In this paper, we will study the aging of the open universe in which $t \to \infty$. Starting from the quantum hyperbolic heat transfer equation, we argue that the Planck time $\tau_P = \left(\frac{\hbar G}{c^5}\right)^{1/2}$ is the border between the time reversible universe $t < \tau_P$ and universe with time arrow for $t > \tau_P$. For time $t \to \infty$, the prevaling thermal process for thermal phenomena in the universe is the diffusion with diffusion constant $D_P = \left(\frac{\hbar G}{c}\right)^{1/2}$, i.e. for $t \to \infty$, $D_P \to \infty$. From formula for D_P, we conclude that for $t \to \infty$, $\hbar \to \infty$ and $c \to 0$. In that case, from formula for fine structure constant α, we obtain $\alpha = $ constant for $t \to \infty$. This result does not exclude the observed very small change of α for redshift $0.5 < z < 3.5$ [8.16, 8.17]. The theory with variable c was considered by J. Magueijo [8.18]. It seems quite interesting that in our scenario, $t \to \infty$, $c \to 0$, $\hbar \to \infty$, the aging universe will be more and more quantum Universe.

The enigma of Planck era, i.e. the event characterized by the Planck time, Planck radius, and Planck mass, is very attractive for speculations. In this section, we discuss the new interpretation of Planck time. We define Planck gas – a gas of massive particles all with masses equal the Planck mass $M_P = \left(\frac{\hbar c}{G}\right)^{1/2}$, and relaxation time for transport process equals the Planck time $\tau_P = \left(\frac{\hbar G}{c^5}\right)^{1/2}$. To the description of a thermal transport process in a Planck gas, we apply the quantum Heaviside heat transport equation (QHH) [8.11]

$$\frac{\lambda_B}{v_h} \frac{\partial^2 T}{\partial t^2} + \frac{\lambda_B}{\lambda} \frac{\partial T}{\partial t} = \frac{\hbar}{M_P} \nabla^2 T. \tag{8.20}$$

In Eq. (8.20), M_P is the Planck mass, λ_B the de Broglie wavelength, and λ mean free path for Planck mass. Equation (8.20) describes the dissipation of the thermal energy induced by a temperature gradient ∇T. Recently, the dissipation of the thermal energy in the cosmological context (e.g., viscosity) was described in the frame of EIT (extended irreversible thermodynamics). With the simple choice for viscous pressure, it is shown that dissipative signals propagate with the light velocity c. Considering that the relaxation time τ is defined as [8.11]

$$\tau = \frac{\hbar}{M_P v_h^2}, \tag{8.21}$$

for thermal wave velocity $v_h = c$, one obtains

$$\tau = \frac{\hbar}{M_P c^2} = \left(\frac{\hbar G}{c^5}\right)^{\frac{1}{2}} = \tau_P, \tag{8.22}$$

i.e. *the relaxation time is equal to the Planck time* τ_P. The gas of massive particles with masses equal to the Planck mass M_P, and relaxation time τ_P we will define as the Planck gas.

According to the results of paper [8.11], we define the quantum of the thermal energy, *heaton*, for the Planck gas as

$$E_h = \hbar\omega_P = \frac{\hbar}{\tau_P} = \left(\frac{\hbar c}{G}\right)^{\frac{1}{2}} c^2 = M_P c^2, \tag{8.23}$$

i.e.

$$E_h = \hbar\omega_P = 10^{19}\,\text{GeV}. \tag{8.24}$$

With formula (8.21), and $v_h = c$ we calculate the mean free path λ, viz.

$$\lambda = v_h \tau_P = c\tau_P = \left(\frac{\hbar G}{c^3}\right)^{\frac{1}{2}}. \tag{8.25}$$

From formula (8.25), we conclude that mean free path for a Planck gas is equal to the Planck radius. For a Planck mass, we can calculate the de Broglie wavelength

$$\lambda_B = \frac{\hbar}{M_P v_h} = \left(\frac{G\hbar}{c^3}\right)^{\frac{1}{2}} = \lambda. \tag{8.26}$$

As it is defined in the book [8.11], (8.26) describes the quantum limit of heat transport. When formulae (8.25) and (8.26) are substituted into (8.20), we obtain

$$\tau_P \frac{\partial^2 T}{\partial t^2} + \frac{\partial T}{\partial t} = \frac{\hbar}{M_P} \nabla^2 T. \tag{8.27}$$

Equation (8.27) is the quantum hyperbolic heat transport equation for a Planck gas. It can be written as

$$\frac{\partial^2 T}{\partial t^2} + \left(\frac{c^5}{\hbar G}\right)^{\frac{1}{2}} \frac{\partial T}{\partial t} = c^2 \nabla^2 T. \tag{8.28}$$

The quantum hyperbolic heat equation (8.28) as a hyperbolic equation sheds light on the time arrow in a Planck gas. When QHT is written in the equivalent form

$$\tau_P \frac{\partial^2 T}{\partial t^2} + \frac{\partial T}{\partial t} = D_P \nabla^2 T, \tag{8.29}$$

where $D_P = \left(\frac{\hbar G}{c}\right)^{\frac{1}{2}}$ is the diffusion coefficient for a Planck gas, then for time period shorter then τ_P we have preserved time reversal for thermal processes, viz.

$$\frac{1}{c^2}\frac{\partial^2 T}{\partial t^2} = \nabla^2 T. \tag{8.30}$$

For the aging of the universe, i.e. for $t \to \infty$, $t \gg \tau_P$ the time reversal symmetry is broken

$$\frac{\partial T}{\partial t} = \left(\frac{\hbar G}{c}\right)^{\frac{1}{2}} \nabla^2 T. \tag{8.31}$$

These new properties of (8.28) open up new possibilities for the interpretation of the Planck time. Before τ_P, thermal processes in Planck gas are symmetrical in time. After τ_P, i.e. for $t \to \infty$ the time symmetry is broken. Moreover gravitation is activated after τ_P and this fact creates an arrow of time (8.31).

It is well-known that the equation (8.30) is invariant under Lorentz group transformation whereas equation (8.31) is not. The time border between two processes domination: waves and diffusion is the τ_P. On the other hand, τ_P as the time period is dependent on the observer velocity, i.e. can in principle be different for different observers. The way out that solves the contradictory is to assume that τ_P is invariant under Lorentz transformation. Considering that Planck length is equal

$$L_P = c\tau_P, \tag{8.32}$$

we obtain that L_P, Planck length is invariant under Lorentz transformation. This conclusion is in harmony with the results of the G. Amelino-Camelia paper [8.19].

8.2.2 The Ecosphere and the Value of the Electromagnetic Fine Structure Constant

The existence of extra-solar planets is well established. In the contemporary status of the searching program (e.g. DARWIN space infrared interferometer project), the following categories of extra-solar planets are described: Definite planets (20), possible planets (8), microlensed planets (5), borderline planets (2), dust clump planets (7), and pulsar planets (4); number in parentheses denotes the number of planet.[1] It is well known, that round the Sun the habitable zone Ecosphere exists. Within the Sun Ecosphere are Venus, Earth, and Mars and Sun. It will be interesting to calculate the Ecosphere radius for "average" star with mass $M_s = a_G^{3/2} m_p$ (a_G = the gravity fine structure constant and m_p = proton mass).

To that aim, we investigate the possibility of the calculations of the planet orbit radii as the function of the fine structure constant α. We argue that the Ecosphere is defined as the part of space around the star that can be calculated assuming the present-day value of the electromagnetic fine structure constant. Considering the existence of the "niche" for fine structure constant, we calculate the niche for planet orbit radii and obtain $R_{\text{rel}} = [R(\alpha)]/[R(\alpha = 1/137)] = 0.5 - 1.5$ where R_{rel} denotes the relative orbit radii. In the case of the Solar system in Ecosphere, we find out the orbits

---

[1] Data taken from http://art.star.rl.ac.uk/darwin/planets.

of Venus, Earth, and Mars. Considering the agreement of the calculation with the Ecosphere radius for Solar system, we argue that our model for habitable zone can be applied to other planet systems also.

In paper [8.20], the quantum heat transport on the atomic, nuclear, and quark scale was discussed and the characteristic time scales were obtained. For atomic scale:

$$\tau_a = \frac{\hbar}{m_e \alpha^2 c^2}, \tag{8.33}$$

where m_e is electron mass, α is the electromagnetic fine structure constant, and c is the light velocity. For nuclear scale:

$$\tau_n = \frac{\hbar}{m_n (\alpha^s)^2 c^2}, \tag{8.34}$$

where m_n denotes nucleon mass, and $\alpha^s \sim 0.15$ is the coupling constant for strong interactions. In the case of free quark gas (if it exists!):

$$\tau_q = \frac{\hbar}{m_q (\alpha_s^q)^2 c^2} \tag{8.35}$$

and α_s^q, m_q are the fine structure constant for quark-quark interaction and quark mass, respectively.

The atomic time scale, τ_a, is proportional to the "atomic year", T_a, viz.:

$$\tau_a = \frac{\hbar}{m_e \alpha^2 c^2} \sim \frac{a_{\mathrm{B}}}{\alpha \, c} = T_a, \tag{8.36}$$

where a_{B} is the Bohr radius.

It is quite interesting to observe that the "coincidence" holds:

$$A \, T_a \sim T_{\mathrm{Earth}}, \tag{8.37}$$

$$A \, m_p = 1 \, \mathrm{g},$$

where A is the Avogadro number, $A = 6.02 \, 10^{23}$, $m_p = 1.66 \, 10^{-27}$ kg is the proton mass, and T_{Earth} denotes Earth year (in seconds).

From Kepler law, the relation $T_{\mathrm{Earth}} \to R_{\mathrm{Earth}}$ can be concluded:

$$T_{\mathrm{Earth}}^2 = \left(\frac{2 m_{\mathrm{Earth}} \pi}{(-m_{\mathrm{Earth}} K)^{\frac{1}{2}}} \right)^2 R_{\mathrm{Earth}}^3. \tag{8.38}$$

In formula (8.38), we approximate Earth orbit as the circle with radius R_{Earth}, m_{Earth} is Earth mass, K is equal:

$$K = -G m_{\mathrm{Earth}} M, \tag{8.39}$$

where G is gravity constant and M denotes the mass of the central body (the Sun) that creates gravity forces. In the following, we approximate the M mass of the central body by the mass of the "average" star [8.21, 8.22]

$$M \cong a_G^{-\frac{3}{2}} m_p = N m_p, \qquad (8.40)$$

where $N = a_G^{-3/2}$ is the Landau-Chandrasekhar number, and a_G denotes the fine structure constant for gravity force. Comparing formulae (8.37) and (8.38) one obtains:

$$R^{\frac{3}{2}} = \frac{A \hbar c}{2 \pi \alpha^2} \frac{1}{m_e c^2} \left(\frac{M_{pl}}{m_p} \right)^{\frac{1}{2}} \left(\frac{\hbar c}{m_p c^2} \right)^{\frac{1}{2}}. \qquad (8.41)$$

In formula (8.41), for planet radius, we omit the subscript "Earth" because the radius does not depend on the planet mass. The R denotes the planet orbit radius for average star with mass described by formula (8.40). The planet radius depends only on the three fundamental constants of Nature: $G, \hbar, c$. The mass $M_{pl} = (\frac{\hbar c}{G})^{1/2}$ is the Planck mass.

Considering formula (8.37), $A\, m_p = 1\,\text{g}$ the planet radius (8.41) can be formulated in more "elegant" form:

$$R^{\frac{3}{2}} = \frac{\hbar c}{m_p \alpha^2} \frac{1}{m_e c^2} \left(\frac{M_{pl}}{m_p} \right)^{\frac{1}{2}} \left(\frac{\hbar c}{m_p c^2} \right)^{\frac{1}{2}}. \qquad (8.42)$$

The dependence of R on α is quite interesting. It is well-known that grand unified theories allow very sharp limits to be placed on the possible values of the fine structure constant in a cognizable universe. The possibility of doing physics on the background space-time at the unification energy and the existence of stars made of protons and neutrons endorse α in the niche [8.23]:

$$\frac{1}{180} \leq \alpha \leq \frac{1}{85} \qquad (8.43)$$

or [8.24]

$$\frac{1}{195} \leq \alpha \leq \frac{1}{114}. \qquad (8.44)$$

It is interesting to observe that one obtains the niche for planet radii – the Ecosphere, which is the consequence of formulae (8.43) and (8.44). The Ecosphere spans from $R_{rel} \sim 0.5$ to $R_{rel} \sim 1.5$. In the case of the Solar system, in this niche we find only the orbits of Venus, Earth and on the border of the Ecosphere: Mars.

Considering the agreement of the calculations present in this paper with the habitable zone for Sun, we argue that our model for Ecosphere can be applied to other planet systems (other "worlds") also.

8.2.3 Inconstancy of the Fine Structure Constant, α?

The contemporary observational method can compare the value of fine structure constant $\alpha = \frac{1}{137}$ in different ages of the universe [8.16, 8.17]. For the sources lying between redshifts 0.5 and 3.5 as a whole, the observed shifts is

$$\frac{\Delta\alpha}{\alpha} = \frac{[\alpha(z) - \alpha(\text{now})]}{\alpha(\text{now})} = (-0.72 \pm 0.18)10^{-5}. \qquad (8.45)$$

If one converts this into a rate of charge of α with time, it amounts to about

$$\frac{(\text{rate of change of } \alpha)}{(\text{current value of } \alpha)} = 5 \cdot 10^{-16} \text{per year}. \qquad (8.46)$$

One of the authors of the papers [8.16, 8.17] once wrote on the constants of nature [8.25]:

> There is something attractive about permanence. We feel instinctively that things that have remained unchanged for centuries must posses some attribute that is intrinsically good. . . . And despite the constant flux of changing events, we feel that the world possesses some invariant bedrock where general aspect is the same.

I still share this point of view and assume α is constant through the evolution of universe. But the aging of the Universe means the transition from wavy motion to the diffusion, i.e. the diffusion is dominant for $t \to \infty$. The growing influence of diffusion term in equation (8.21) means:

$$D_P = \left(\frac{\hbar G}{c}\right)^{\frac{1}{2}} \to \infty, \qquad \text{for} \qquad t \to \infty$$

i.e. (when G = constant) $\hbar \to \infty$, $c \to 0$ for $t \to \infty$. In this scenario, $\alpha = (e^2)/(\hbar c)$ can be constant through the life of the universe – our Universe.

One of the conclusions is that for $t \to \infty$, $c \to 0$ is in harmony with new results of João Magueijo [8.18] on the varying of the light speed. It is interesting to observe that in our scenario, i.e. $c \to 0$, $\hbar \to \infty$ for $t \to \infty$ the aging Universe will be more and more quantum Universe with prevailing quantum effects over the classical behavior. Who knows?

References

8.1. M. Kozlowski, J. Marciak-Kozlowska: Lasers Eng. **9**, 365 (1999)
8.2. M. Niering et al.: Phys. Rev. Lett. **84**, 5496 (2000)
8.3. A. Huber et al.: Phys. Rev. Lett. **80**, 468 (1998)
8.4. S.G. Karshenboim et al.: JETP **79**, 230 (1994)
8.5. P.J. Mohr et al.: Rev. Mod. Phys. **72**, 351 (2000)
8.6. S.G. Karshenboim: Can. J. Phys. **78**, 639 (2000)
8.7. L. de la Peña, M. Cetto: *The Quantum Dice: An Introduction to Stochastic Electrodynamics* (Kluwer, 1996)
8.8. B. Haisch, A. Rueda: Phys. Lett. **A268**, (2000)
8.9. H.E. Puthoff.: Phys. Rev. **D35**, 3266 (1987)
8.10. H.E. Puthoff: Phys. Rev. **A39**, 2333 (1989)
8.11. M. Kozlowski, J. Marciak-Kozlowska: *From Quarks to Bulk Matter* (Hadronic Press, 2001)
8.12. R.P. Feynman, R.B. Leighton, M. Sands: *The Feynman Lecture on Physics* (Addison-Wesley, Reading, MA 1963)
8.13. T.H. Boyer: Phys. Rev. **D11**, 790–809 (1975)
8.14. J.N. Islam: *An Introduction to Mathematical Cosmology* (CUP 2002)
8.15. F.J. Dyson: Rev. Mod. Phys. **51**, 447 (1979)
8.16. J.K. Webb et al.: Phys. Rev. Lett. **82**, 884 (1999)
8.17. J.K. Webb et al.: Phys. Rev. Lett. **87**, 091301 (2001)
8.18. J. Magueijo: *Faster than the Speed of Light* (Perseus Publishing 2003)
8.19. G. Amelino-Camelia: Phys. Lett. B **510**, 255 (2001)
8.20. J. Marciak-Kozlowska, M. Kozlowski: Found. Phys. Lett. **9**, 285 (1996)
8.21. L.D. Landau, E.M. Lifshiftz: *Statistical Physics*, 2nd ed. (Pergamon Press, Oxford 1997)
8.22. S. Chandrasekhar, Mon. Mot. R. Astron. Soc. **95**, 201 (1935)
8.23. J.D. Barrow, F.J. Tipler: *The Antropic Cosmological Principle* (Oxford University Press, Oxford 1986)
8.24. M. Kozlowski: *Physics Essays*, **7**, (1994), p. 261.
8.25. J.D. Barrow: *Theories of Everything* (Fawcett Columbine, New York 1991) p 117
8.26. S.G. Karshenboim: Can. J. Phys. **77**, 241 (1999)

9

Epilogue: The Emergence of Quantum Dynamics in a Classical World

9.1 Key Questions

There were three major crises in physics in the 20th century. In each case, two well-established theories were found to be incompatible, either because they were based on contradictory assumptions about the workings of the physical world or because they led to physically untenable conclusions. The first was the conflict between Newtonian gravity and special relativity, which was resolved by Einstein's theory of general relativity.

The second arose from tension between thermodynamics and electromagnetism, which led to the development of quantum theory. In both of these cases, the crisis was resolved not by small modification of one of the theories but rather by an entirely new fundamental theory that introduced a new framework and made some old concepts obsolete. Over time, both of these theories passed stringent experimental tests and now form the cornerstone of modern physics. Unfortunately, however, general relativity and quantum theory are themselves mutually incompatible, presenting physicists with a third crisis.

Unlike the theory of relativity, the other great idea that shaped physical notions at the same time, quantum mechanics does far more than modify Newton's equations of motion. Whereas relativity redefines the concepts of space and time in terms of the observer, quantum mechanics denies an aspect of reality to system properties (such as position and momentum) until they are measured.

The basic question for understanding the classical-quantum borderline is: how come the world appears to be classical when the fundamental theory describing it is manifestly not so? One of the reasons the quantum-to-classical transition took so long to come under serious investigation may be that it was confused with the measurement problem. In fact, the problem of assigning intrinsic reality to properties of individual quantum systems gave rise to a purely statistical interpretation of quantum mechanics. In this view, *quantum laws apply only to ensembles of identically prepared systems.*

As experimental technology progressed to the point at which single electron transistor could be measured with precision, the facade of ensemble statistics could no longer hide the reality of the counter-classical nature of quantum mechanics and quantum technology. In particular, a vast array of quantum features, such as interference and tunneling, can be seen as everyday occurrences in quantum technology.

9.2 Schrödinger-Newton Wave Mechanics: The Model

When M. Planck made the first quantum discovery, he noted an interesting fact [9.1]. The speed of light, Newton's gravity constant, and Planck's constant clearly reflect fundamental properties of the world. From them it is possible to derive the characteristic mass M_P, length L_P, and time T_P with approximate values:

$$L_P = 10^{-35} \text{m},$$

$$T_P = 10^{-43} \text{s},$$

$$M_P = 10^{-5} \text{g}.$$

Nowadays, much of cosmology is concerned with "interface" of gravity and quantum mechanics.

After the *Alpha* moment – the spark in eternity [9.1] the space and time were created by "Intelligent Design" [9.2] at $t = T_P$. The enormous numbers efforts of the physicists, mathematicians and philosophers investigate the *Alpha* moment. Scholars seriously discuss the *Alpha* moment by all possible means: theological and physico-mathematical with growing complexity of theories. The most important result of these investigations is the anthropic principle and Intelligent Design Theory (ID) [9.9].

In this section, we investigate the very simple question: how gravity can modify the quantum mechanics, i.e. the nonrelativistic Schrödinger equation (SE). We argue that SE with relaxation term describes properly the quantum behavior of particle with mass $m < M_\mathrm{P}$ and contains the part that can be interpreted as the pilot wave equation. For $m \to M_\mathrm{P}$, the solution of the SE represents the *strings* with mass M_P.

9.2.1 Generalized Fourier Law

The thermal history of the system (heated gas container, semiconductor, or Universe) can be described by the generalized Fourier equation [9.3]–[9.5]

$$q(t) = \int_{-\infty}^{t} \underbrace{K(t - t')}_{\text{thermal history}} \underbrace{\nabla T(t')}_{\text{diffusion}} dt' . \tag{9.1}$$

In equation (9.1) $q(t)$ is the density of the energy flux, T is the temperature of the system, and $K(t - t')$ is the thermal memory of the system

$$K(t - t') = \frac{K}{\tau} \exp\left[-\frac{(t - t')}{\tau}\right] , \tag{9.2}$$

where K is constant, and τ denotes the relaxation time.

As was shown in [9.3]–[9.5]

$$K(t - t') = \begin{cases} K \lim_{t_0 \to 0} \delta(t - t' - t_0) & \text{diffusion} \\ K = \text{constant} & \text{wave} \\ \dfrac{K}{\tau} \exp\left[-\dfrac{(t - t')}{\tau}\right] & \text{damped wave or hyper-} \\ & \text{bolic diffusion} . \end{cases}$$

The damped wave or hyperbolic diffusion equation can be written as:

$$\frac{\partial^2 T}{\partial t^2} + \frac{1}{\tau} \frac{\partial T}{\partial t} = \frac{D_T}{\tau} \nabla^2 T . \tag{9.3}$$

For $\tau \to 0$, (9.3) is the Fourier thermal equation

$$\frac{\partial T}{\partial t} = D_T \nabla^2 T \tag{9.4}$$

and D_T is the thermal diffusion coefficient. The systems with very short relaxation time have very short memory. On the other hand, for $\tau \to \infty$, (9.3) has the form of the thermal wave (undamped) equation, or *ballistic* thermal equation. In solid-state physics, the *ballistic* phonons or electrons are those for which $\tau \to \infty$. The experiments with *ballistic* phonons or electrons demonstrate the existence of the *wave motion* on the lattice scale or on the electron gas scale.

$$\frac{\partial^2 T}{\partial t^2} = \frac{D_T}{\tau} \nabla^2 T . \tag{9.5}$$

For the systems with very long memory, (9.3) is time symmetric equation with no arrow of time, for Eq. (9.5) does not change the shape when $t \to -t$.

In equation (9.3), we define:

$$v = \left(\frac{D_T}{\tau}\right)^{\frac{1}{2}},$$

(9.6)

velocity of thermal wave propagation and

$$\lambda = v\tau$$

(9.7)

where λ is the mean free path of the heat carriers. With formula (9.6), equation (9.3) can be written as

$$\frac{1}{v^2}\frac{\partial^2 T}{\partial t^2} + \frac{1}{\tau v^2}\frac{\partial T}{\partial t} = \nabla^2 T.$$

(9.8)

9.2.2 Damped Wave Equation, Thermal Carriers in Potential Well, V

From the mathematical point of view, equation

$$\frac{1}{v^2}\frac{\partial^2 T}{\partial t^2} + \frac{1}{D}\frac{\partial T}{\partial t} = \nabla^2 T$$

is the hyperbolic partial differential equation (PDE). On the other hand, Fourier equation

$$\frac{1}{D}\frac{\partial T}{\partial t} = \nabla^2 T$$

(9.9)

and Schrödinger equation

$$i\hbar\frac{\partial \Psi}{\partial t} = -\frac{\hbar^2}{2m}\nabla^2 \Psi$$

(9.10)

are the parabolic equations. Formally with substitutions

$$T \leftrightarrow \Psi, \quad t \leftrightarrow it$$

(9.11)

Fourier equation (9.9) can be written as

$$i\hbar\frac{\partial \Psi}{\partial t} = -D\hbar\nabla^2 \Psi$$

(9.12)

and by comparison with Schrödinger equation, one obtains

$$D_T\hbar = \frac{\hbar^2}{2m}$$

(9.13)

and

$$D_T = \frac{\hbar}{2m} \,. \tag{9.14}$$

Considering that $D_T = \tau v^2$ (9.6), we obtain from (9.14)

$$\tau = \frac{\hbar}{2mv^2} \,. \tag{9.15}$$

Formula (9.15) describes the relaxation time for quantum thermal precesses.

Starting with Schrödinger equation for particle with mass m in potential V:

$$i\hbar\frac{\partial\Psi}{\partial t} = -\frac{\hbar^2}{2m}\nabla^2\Psi + V\Psi \tag{9.16}$$

and performing the substitution (9.11), one obtains

$$\hbar\frac{\partial T}{\partial t} = \frac{\hbar^2}{2m}\nabla^2 T - VT \tag{9.17}$$

and

$$\frac{\partial T}{\partial t} = \frac{\hbar}{2m}\nabla^2 T - \frac{V}{\hbar}T \,. \tag{9.18}$$

Equation (9.18) is Fourier equation (parabolic PDE) for $\tau = 0$. For $\tau \neq 0$ we obtain

$$\tau\frac{\partial^2 T}{\partial t^2} + \frac{\partial T}{\partial t} + \frac{V}{\hbar}T = \frac{\hbar}{2m}\nabla^2 T \,, \tag{9.19}$$

$$\tau = \frac{\hbar}{2mv^2} \tag{9.20}$$

or

$$\frac{1}{v^2}\frac{\partial^2 T}{\partial t^2} + \frac{2m}{\hbar}\frac{\partial T}{\partial t} + \frac{2mV}{\hbar^2}T = \nabla^2 T \,.$$

9.2.3 Model Schrödinger Equation

With the substitution (9.11), equation (9.19) can be written as

$$i\hbar\frac{\partial\Psi}{\partial t} = V\Psi - \frac{\hbar^2}{2m}\nabla^2\Psi - \tau\hbar\frac{\partial^2\Psi}{\partial t^2} \,. \tag{9.21}$$

The new term, relaxation time

$$\tau\hbar\frac{\partial^2\Psi}{\partial t^2} \tag{9.22}$$

describes the interaction of the particle with mass m with spacetime. The relaxation time τ can be calculated as:

$$\frac{1}{\tau} = \frac{1}{\tau_{e-p}} + \cdots + \frac{1}{\tau_{\text{Planck}}} \qquad (9.23)$$

where, for example τ_{e-p} denotes the scattering of the particle m on the electron-positron pair ($\tau_{e-p} \sim 10^{-17}$ s), and the shortest relaxation time τ_{Planck} is the Planck time ($\tau_{\text{Planck}} \sim 10^{-43}$ s).

From equation (9.23), we conclude that $\tau \approx \tau_{\text{Planck}}$ and (9.21) can be written as

$$i\hbar \frac{\partial \Psi}{\partial t} = V\Psi - \frac{\hbar^2}{2m}\nabla^2\Psi - \tau_{\text{Planck}}\hbar \frac{\partial^2 \Psi}{\partial t^2}, \qquad (9.24)$$

where

$$\tau_{\text{Planck}} = \frac{1}{2}\left(\frac{\hbar G}{c^5}\right)^{\frac{1}{2}} = \frac{\hbar}{2M_{\text{P}}c^2}. \qquad (9.25)$$

In formula (9.25), M_{P} is the mass Planck. Considering (9.25), (9.24) can be written as

$$i\hbar \frac{\partial \Psi}{\partial t} = -\frac{\hbar^2}{2m}\nabla^2\Psi + V\Psi - \frac{\hbar^2}{2M_{\text{P}}}\nabla^2\Psi + \frac{\hbar^2}{2M_{\text{P}}}\nabla^2\Psi - \frac{\hbar^2}{2M_{\text{P}}c^2}\frac{\partial^2\Psi}{\partial t^2}. \qquad (9.26)$$

The last two terms in (9.6) can be defined as the *Bohmian* pilot wave

$$\frac{\hbar^2}{2M_{\text{P}}}\nabla^2\Psi - \frac{\hbar^2}{2M_{\text{P}}c^2}\frac{\partial^2\Psi}{\partial t^2} = 0 \qquad (9.27)$$

i.e.

$$\nabla^2\Psi - \frac{1}{c^2}\frac{\partial^2\Psi}{\partial t^2} = 0. \qquad (9.28)$$

It is interesting to observe that pilot wave Ψ does not depend on the mass of the particle. With postulate (9.28), we obtain from equation (9.26)

$$i\hbar \frac{\partial \Psi}{\partial t} = -\frac{\hbar^2}{2m}\nabla^2\Psi + V\Psi - \frac{\hbar^2}{2M_{\text{P}}}\nabla^2\Psi \qquad (9.29)$$

and simultaneously

$$\frac{\hbar^2}{2M_{\text{P}}}\nabla^2\Psi - \frac{\hbar^2}{2M_{\text{P}}c^2}\frac{\partial^2\Psi}{\partial t^2} = 0. \qquad (9.30)$$

In the operator, form (9.9) and (9.20) can be written as

$$\hat{E} = \frac{\hat{p}^2}{2m} + \frac{1}{2M_{\text{P}}c^2}\hat{E}^2. \qquad (9.31)$$

where $\hat{E}$ and $\hat{p}$ denote the operator for energy and momentum of the particle with mass m. Equation (9.31) is the new dispersion relation for quantum particle with mass m. From (9.31), one can conclude that Schrödinger quantum mechanics is valid for particles with mass $m \ll M_P$.[1] But pilot wave Ψ exists independent of the mass of the particles.

For particles with mass $m \ll M_P$, (9.9) has the form

$$i\hbar\frac{\partial \Psi}{\partial t} = -\frac{\hbar^2}{2m}\nabla^2\Psi + V\Psi\,. \tag{9.32}$$

9.2.4 Schrödinger Equation and the Strings

In the case when $m \approx M_P$, (9.29) can be written as

$$i\hbar\frac{\partial \Psi}{\partial t} = -\frac{\hbar^2}{2M_P}\nabla^2\Psi + V\Psi \tag{9.33}$$

but considering (9.30), one obtains

$$i\hbar\frac{\partial \Psi}{\partial t} = -\frac{\hbar^2}{2M_P c^2}\frac{\partial^2 \Psi}{\partial t^2} + V\Psi \tag{9.34}$$

or

$$\frac{\hbar^2}{2M_P c^2}\frac{\partial^2 \Psi}{\partial t^2} + i\hbar\frac{\partial \Psi}{\partial t} - V\Psi = 0\,. \tag{9.35}$$

We look for the solution of (9.35) in the form

$$\Psi(x,t) = e^{i\omega t}u(x)\,. \tag{9.36}$$

After substitution of formula (9.16) into (9.35), we obtain

$$\frac{\hbar^2}{2M_P c^2}\omega^2 + \omega\hbar + V(x) = 0 \tag{9.37}$$

with the solution

$$\omega_1 = \frac{-M_P c^2 + M_P c^2\sqrt{1 - \frac{2V}{M_P c^2}}}{\hbar}\,, \tag{9.38}$$

$$\omega_2 = \frac{-M_P c^2 - M_P c^2\sqrt{1 - \frac{2V}{M_P c^2}}}{\hbar}\,,$$

for $\frac{M_P c^2}{2} > V$ and

$$\omega_1 = \frac{-M_P c^2 + iM_P c^2\sqrt{\frac{2V}{M_P c^2} - 1}}{\hbar}\,, \tag{9.39}$$

$$\omega_2 = \frac{-M_P c^2 - iM_P c^2\sqrt{\frac{2V}{M_P c^2} - 1}}{\hbar}$$

---

[1] The observation of cosmic ray (CR) events with energy $E \geq 10^{11}$ suggests that according to formula (9.31), the deviations from Schrödinger quantum mechanics can be experimentally verified. Compare: M. Drees: http://lanl.arxiv.org/hep-ph/0304030.

212 9 Epilogue: The Emergence of Quantum Dynamics in a Classical World

for $\frac{M_P c^2}{2} < V$.

Both formulae (9.38) and (9.39) describe the string oscillation, formula (9.27) damped oscillation, and formula (9.28) overdamped string oscillation.

D. Bohm presented the pilot wave theory in 1952, and de Broglie had presented a similar theory in 1923 [9.8]. It was rejected in the 1950s, and the initial rejection had nothing to do with Bohm's later work.

There is always the possibility that the pilot wave has a primitive, mind-like property. That's how Bohm described it. We can say that all the particles in Universe and even Universe have their own *pilot waves*, their own information. In this section we discuss in a very simple nonrelativistic way the possible extension of the Schrödinger equation with relaxation process included. As the first approximation, this leads us to the inclusion of the Planck time, i.e. gravity to the quantum description of the processes in spacetime. The relaxation time τ_{Planck} is the decoherence time [9.6] or the Ehrenfest time [9.7] and describes the collapse of the pilot wave after the interaction with the apparatus.

The new field of investigation, attophysics (or physics at the attosecond frontier), opens the new possibility for the study of the classical-quantum borderline. At these time scales, chemistry is essentially frozen in time, so the only dynamics to be studied are those of electrons, as they are much lighter and faster than nuclei. The attosecond laser pulses will show the quite distinct world-quantum world. In this world, the devices built by quantum engineers will operate according to laws deduced from the Schrödinger equation – from the modified Schrödinger equation, I hope.

References

9.1. J. Barbour: *The End of Time* (Oxford University Press 2000)
9.2. J.F. Addicot: *Ohio State Law Journal* **64**, 125 (2003)
9.3. M. Kozlowski, J. Marciak-Kozlowska: Found. Phys. Lett. **10**, 295 (1997)
9.4. M. Kozlowski, J. Marciak-Kozlowska: Found. Phys. Lett. **10**, 599 (1997)
9.5. M. Kozlowski: J. Marciak-Kozlowska: Found. Phys. Lett. **12**, 93 (1999)
9.6. T. Geszti: http://lanl.arxiv.org/quant-ph/0401086
9.7. G.P. Berman et al.: http://lanl.arxiv.org/quant-ph/0401038
9.8. L. de Broglie: Comptes Rendus **vol. 177**, 507 (1923)
9.9. M. Kozlowski, J. Marciak-Kozlowska: Hadronic Journal **28**, 209 (2005)

A

Elliptic, Parabolic, and Hyperbolic Equations

The hyperbolic heat transport equation

$$\frac{1}{v^2}\frac{\partial^2 T}{\partial t^2} + \frac{m}{\hbar}\frac{\partial T}{\partial t} + \frac{2Vm}{\hbar^2}T - \frac{\partial^2 T}{\partial x^2} = 0 \qquad (A.1)$$

is the partial two-dimensional differential equation (PDE). According to the classification of the PDE, QHT is the hyperbolic PDE. To show this, let us consider the general form of PDE, with only two independent variables ξ and η

$$\left(A\frac{\partial^2}{\partial \xi^2} + B\frac{\partial^2}{\partial \xi \eta} + C\frac{\partial^2}{\partial \eta^2}\right)\Psi(\xi,\eta) = \left(\text{function of } \xi, \eta, \Psi, \frac{\partial \Psi}{\partial \xi}, \frac{\partial \Psi}{\partial \eta}\right). \quad (A.2)$$

Then the equation is called

$$\text{elliptic if } B^2 - 4AC < 0,$$

$$\text{parabolic if } B^2 - 4AC = 0, \qquad (A.3)$$

$$\text{hyperbolic if } B^2 - 4AC > 0.$$

For Eq. (A.1), we have

$$A = \frac{1}{v}, \qquad B = 0, \qquad C = -1,$$

i.e. $B^2 - 4AC = 4\,v^{-1} > 0$ and according to (A.3), Eq. (A.1) is the hyperbolic equation.

On the other hand, for the Fourier equation with potential term

$$\frac{m}{\hbar}\frac{\partial T}{\partial t} + \frac{2Vm}{\hbar^2}T - \frac{\partial^2 T}{\partial x^2} = 0 \qquad (A.4)$$

we have

$$A = 0, \qquad B = 0, \qquad C = -1,$$

i.e. $B^2 - 4AC = 0$ and Fourier equation is the parabolic PDE. One can say that for heat transfer, described by Fourier equation, the velocity of heat propagation v is infinite and $A = 0$.

Index

Springer Series in
OPTICAL SCIENCES